Studies in Systems, Decision and Control

Volume 345

The series "Studies in Systems, Decision and Control" (SSDC) covers both new developments and advances, as well as the state of the art, in the various areas of broadly perceived systems, decision making and control–quickly, up to date and with a high quality. The intent is to cover the theory, applications, and perspectives on the state of the art and future developments relevant to systems, decision making, control, complex processes and related areas, as embedded in the fields of engineering, computer science, physics, economics, social and life sciences, as well as the paradigms and methodologies behind them. The series contains monographs, textbooks, lecture notes and edited volumes in systems, decision making and control spanning the areas of Cyber-Physical Systems, Autonomous Systems, Sensor Networks, Control Systems, Energy Systems, Automotive Systems, Biological Systems, Vehicular Networking and Connected Vehicles, Aerospace Systems, Automation, Manufacturing, Smart Grids, Nonlinear Systems, Power Systems, Robotics, Social Systems, Economic Systems and other. Of particular value to both the contributors and the readership are the short publication timeframe and the world-wide distribution and exposure which enable both a wide and rapid dissemination of research output.

Indexed by SCOPUS, DBLP, WTI Frankfurt eG, zbMATH, SCImago.

All books published in the series are submitted for consideration in Web of Science.

More information about this series at http://www.springer.com/series/13304

Pavol Božek · Yury Nikitin · Tibor Krenický

Diagnostics of Mechatronic Systems

Springer

Pavol Božek
Faculty of Materials Science
and Technology
Institute of Production Technologies
Slovak University of Technology
Bratislava, Slovakia

Yury Nikitin
Department of Mechatronic Systems
Kalashnikov Izhevsk State Technical
University
Izhevsk, Russia

Tibor Krenický
Department of Design and Monitoring
of Technical Systems
Faculty of Manufacturing Technologies
with a seat in Prešov
Technical University of Košice
Prešov, Slovakia

ISSN 2198-4182 ISSN 2198-4190 (electronic)
Studies in Systems, Decision and Control
ISBN 978-3-030-67054-2 ISBN 978-3-030-67055-9 (eBook)
https://doi.org/10.1007/978-3-030-67055-9

This Springer imprint is published by the registered company Springer Nature Switzerland AG
The registered company address is: Gewerbestrasse 11, 6330 Cham, Switzerland

Reviewers

Andrey Ivanovich Abramov, Assoc. Prof. Ph.D., Izhevsk State Technical University M. T. Kalashnikov, Izhevsk, Russia

Witold Biały, Dr. hab. inż. Prof. PŚ, Politechnika Śląska, Institute of Industrial Engineering, Zabrze, Poland

Ivan Kuric, Prof. Dr. Ing., University of Žilina, Department of Automation and Production, Faculty of Mechanical Engineering, Žilina, Slovakia

About This Book

This scientific monograph defines the basic terms and definitions of diagnostics related to mechatronic systems and describes methods and devices for diagnostics of mechatronic systems. Algorithms for diagnostics of mechatronic systems are also mentioned. The monograph develops models of systems for the diagnosis of mechatronic objects, examples of neural network development for bearing diagnostics, examples of diagnostic devices, mechatronic dynamic modules based on a fuzzy inference system. Schemes of hardware diagnostics of mechatronic systems are also proposed. Finally, an example of multicriteria optimization of diagnostic systems is presented and solved. The appendixes provide example source codes for machine diagnostics programs with CNC programs for the fuzzy output system.

The monograph is intended for professionals focusing on production equipment and systems, automated production systems, and mechatronics.

The study was mainly carried out in cooperation between the Slovak University of Technology in Bratislava and the Kalashnikov Izhevsk State Technical University with financial support from the projects KEGA MŠ SR No. 015 STU-4/2018 entitled *specialized laboratory with the support of MM textbook for teaching the subject "Design and operation of production systems"* and the operational program research and innovation supporting the project *Research of advanced methods of intelligent information processing*, ITMS Code FP313010T570, co-financed by the European Regional Development Fund. This work was also supported by the Slovak Research and Development Agency under contract no. APVV-18-0316 and by the grant agency VEGA within project No. 1/0393/18 entitled *"Research of Methods for Modeling and Compensation of Hysteresis in Pneumatic Artificial Muscles and PAM-actuated Mechanisms to Improve the Control Performance Using Computational Intelligence."*

Introduction

Modern mechatronic systems use high degree of automation. Mechatronics technologies and the development of microsystem technology belong to the critical technologies worldwide. The inability to build mechatronic systems leads to significant economic losses and can be dangerous to human life and health. Effective diagnostic systems are needed for the early detection of errors in mechatronic systems, for the organization of repairs, and for the assessment of the quality of performed repairs.

Currently, great attention is paid to the diagnostics of complex technical systems. Diagnostics is widely used in aerospace, air, rail, road, water, and pipeline transportation, energy, metallurgy, metalworking, fuels, mining, chemical, paper industry, etc. The use of technical diagnostic systems can significantly reduce need for repair complexity and time, increase productivity, and reduce waste, that is, supported by the increasing number of publications, dissertations, specialized international conferences, and exhibitions worldwide, particularly developing in knowledge-oriented countries that dictate the pace in the field through patents and publications in prestigious journals or proceedings.

At present, implementing principles of the Industry 4.0, it is possible and necessary to use a wide range of sensors, powerful computing systems and artificial intelligence methods based on neural networks, fuzzy inference systems, and genetic algorithms in diagnostic systems for mechatronic objects. Unfortunately, the practical use of artificial intelligence in the diagnosis of mechatronic systems is still quite limited, with a certain level of results privatization and local focus of the issue.

The monograph comprises selected fundamental principles supplemented by results of studies performed with the support of an analytical target project KEGA No. 015STU-4/2018 entitled specialized laboratory with the support of MM textbook for teaching the subject "Design and operation of production systems" for Slovak Technical University Bratislava, which deals with building a specialized laboratory for diagnostics and reliability of the technical system, university textbook, and multimedia teaching application for the subject "Design and operation of production systems." The main mission of the book is to increase the level of the pedagogical process of universities with the support of video sequences, animations, images, and other multimedia conditions for professional subjects. Moreover, authors would like to thank for support of the Slovak Research and Development Agency under contract no. APVV-18-0316.

Moreover, the monograph is an output of a scientific team school led by Prof. Ivan Vasilyevich Abramov, long-year rector of the Kalashnikov Izhevsk State Technical University. Professor Abramov has created original methodology and theory of diagnostics and belongs among the notable personalities that significantly influenced this field of research. The scientific school of Prof. Abramov has trained numerous engineers and researchers in the field of mechatronic systems. His initiatives are also supported within international collaboration, particularly with specialists from the Slovak University of Technology in Bratislava. For the original results of the long-time effort associated with his name, Prof. Abramov deserves deepest respect and gratitude of many, including authors of this monograph.

Contents

List of Figures

List of Tables

Chapter 1
The Basics Characteristics of Elements Reliability

1.1 Core Concepts

Product ***quality*** is a set of characteristics, expressing the ability to perform the intended functions. Simultaneously, the economic indicators of the product are considered, its equipment with accessories, spare parts, etc., as well as the assumptions that are created by the manufacturer for the provision of services related to the use of the product [17, 18].

Reliability is a general property of the object, consisting of the ability to perform the required functions while maintaining the values of the specified operating indicators within the given limits and in time according to the specified technical conditions [27]. Reliability is one of the most important groups of product quality features.

Following partial properties are included [21, 24, 25]:

- reliability,
- durability,
- sustainability,
- repairability,
- operability,
- security.

Reliability is the ability of an object to continuously perform required functions for a specified period of time under specified conditions.

Durability is the ability of the object to perform the required functions until the marginal state is reached while a specified maintenance and repairs system is maintained.

Sustainability is a property of the object consisting of the ability to prevent and determine the causes of its errors and to eliminate their consequences by specified maintenance and repair system.

Repairability is a property of the object that includes the ability to determine the causes of its errors and eliminate their consequences by repair.

P. Božek et al., *Diagnostics of Mechatronic Systems*, Studies in Systems, Decision and Control 345, https://doi.org/10.1007/978-3-030-67055-9_1

Operability is a complex property including the reliability and repairability of the object under operating conditions.

There are several definitions of the term reliability in the literature dealing with reliability issues. Upon closer examination, we find that there are two tendencies: quantitative and qualitative definition of reliability. The quantitative—numerical expression of reliability is defined by numerical characteristics. When solving basic tasks in the field of reliability, two mutually exclusive product states are considered:

error-free operation status "1" and error downtime status "0".

1.2 Terms and Definitions of Mechatronics and Diagnostics

Mechatronics is a relatively new trend in science and technology, and therefore there are many terms and definitions. The following definitions of mechatronic and mechatronic systems are given in the educational standard of higher education in the field of education "Mechatronics and robotics", approved in 2009.

Mechatronics is an area of science and technology, which is based on the system integration of nodes of precision mechanics, environmental sensors and the object itself, energy sources, actuators, amplifiers, computing devices (computers and microprocessors).

Mechatronic system is the only complex of electromechanical, electrohydraulic, electronic elements and computing technology, among which there is a constantly dynamically changing energy and information, combined with a common automatic control system with elements of artificial intelligence.

Below are listed some of the definitions of mechatronics from international literature.

Mechatronics is integration of microprocessor control systems, electrical and mechanical systems.

Mechatronics is an interdisciplinary technological department, which represents the next generation of machines, robots, intelligent mechanisms performing tasks in various industries.

Mechatronics is the synergistic integration of mechanics with electronics and intelligent control algorithms [15].

Currently, mechatronic modules and systems are widely used in these areas:

- machine tools and automatic equipment,
- robotics,
- aviation,
- space technology,
- military equipment,
- vehicles,
- office and household appliances,
- elements of computing technology,
- medical equipment.

In mechatronic systems, adaptive or intelligent control is used, in which it is necessary to have information about the external environment and the state of all mechatronic modules or nodes. To control the diagnostic process of mechatronic systems, it is necessary to know which modules and elements this system consists of and what methods and tools are used to diagnose these modules and elements [8, 14, 23]. At present, there is no general theory of diagnostics of mechatronic systems and there is no generally accepted classification of levels of mechatronic modules.

The first level includes a mechatronic unit or mechatronic module (MM). MM is a unified mechatronic object (MO), which has autonomous documentation and is usually designed to perform movements in one coordinate. Examples of MM are machine parts—headstock, rotary table [6]. The modules can be motors, gearboxes, etc. More complex modules—autonomous drives—motor-reducer, motor-wheel, motor-spindle, motor-drum, rotary table [1–3]. The mechatronic unit differs fundamentally from MM in that it is not uniform.

The second level is the unit (machine), which contains several modules designed to implement the specified laws of motion in the conditions of interaction with the external environment [5, 16]. Examples of units are robots, CNC machines, etc.

The third level is a mechatronic system (MS), consisting of several units or aggregates and the number of individual modules, i.e. objects of the same or different lower levels [7, 11, 26]. A system is a set of components that are interconnected in some way: they are subject to a certain relationship, dependence, or pattern that acts as one. MS fully meets this definition as a set of mechanical, electronic, and control components that create a synergistic unity [20].

Examples of MS include flexible manufacturing systems, mobile robots, or modern vehicles [13, 19, 28]. Technical diagnostics is required for the reliable and efficient operation of mechatronic systems [4].

- ***Technical diagnostics*** is the science of recognizing the state of the technical system. The term "diagnosis" comes from the Greek word "diagnosis", which means recognition, definition. The technical diagnostics aims to increase the reliability and resources of technical systems. The tasks of technical diagnostics are the detection of errors, their location, and determining the causes of errors.
- ***Diagnostics*** is an area of knowledge that examines the technical condition of diagnosed objects and the manifestation of technical conditions, develops methods for their determination, as well as the principles of construction and organization of the use of diagnostic systems. When diagnoses are technical subjects, they speak of technical diagnostics.

The subject of technical diagnostics is the product and its components, the technical condition of which is to be determined.

- ***Technical diagnostics*** is the process of determining the technical condition of the diagnosed object with a certain accuracy.
- ***Technical condition of the subject of diagnosis*** is a set of properties of the object, which may change in the process of production or operation, characterized at a certain point in time by the characters determined by the technical documentation

for this object. The parameter of the technical condition is a physical quantity that characterizes the performance or operability of the diagnosed object, which changes during the work.

- ***Diagnostic parameter*** is the diagnosis object parameter, used in a prescribed manner to determine the technical condition of the diagnosis object.
- ***Structural parameter*** is a parameter that directly characterizes the performance of the diagnosed object (wear, play, voltage, etc.).
- ***Measurement*** is determining the value of a physical quantity empirically using special technical means.
- ***Diagnosis reliability*** is the probability that the diagnosis determines the actual technical state of the diagnosed object.
- ***Prediction of the technical condition*** is prediction of changes in the parameters of the technical condition of the diagnosis object in the future.
- ***Work time*** is the duration of the operation of the facility or the amount of work performed during a certain period of time.
- ***Average time to failure*** is the average lifetime value of the repaired product between failures.
- ***Residual source*** is the operating time of the object to be diagnosed until the maximum change of the parameter of its technical condition, starting from the moment of diagnostics.
- ***Reliability*** is the property of an object to perform specified functions, while maintaining the values of specified performance indicators within specified limits, corresponding to specified modes and conditions of use, for the required period or the required operating time. Quantitatively, reliability is assessed by reliability, durability, maintainability, and stability.
- ***Efficiency*** represents the condition of the product, at which at a given moment its main (operational) parameters are within the limits set by the requirements of the technical documentation.
- ***Durability*** is the properties of an object to maintain performance until the condition of the maintenance and repair system occurs.
- ***Testability*** is a property of the product that characterizes its adaptability to be controlled by specified means.
- ***Sustainability*** is properties of the object, which consists in adapting to the prevention and detection of the causes of its errors and elimination of their consequences by performing repairs and maintenance.

Next, the following assumptions will be considered:

- *mutual independence of errors of individual elements,*
- *exponential probability distribution of errors and repairs,*
- *prompt implementation of repairs characterized by immediate repair in the time of element creation.*

For further use, we will present only the following characteristic quantities and functions.

1.3 The Element Reliability Characteristics

The basic elements or device reliability characteristics are **the probability of error-free operation R(t)** and the **error probability F(t)** defined by Eqs. (1.1) and (1.2). These characteristics take on a different shape depending on the progress of error $\lambda(t)$.

$F_i(t)$ the failure probability of the ith element, wherein:

$$F_i(t) = 1 - \exp(-\lambda \cdot t) \tag{1.1}$$

$R_i(t)$ the error-free operation probability of ith element, wherein:

$$R_i(t) = 1 - F_i(t) = \exp(-\lambda \cdot t) \tag{1.2}$$

Ai (t) the readiness of the repairable ith element at time t (assuming that at time t = 0 it is in normal operation 0, where:

$$A_i(t) = \frac{\mu_i}{\mu_i + \lambda_i} + \frac{\lambda_i}{\mu_i + \lambda_i} e^{-(\mu_i + \lambda_i)t}. \tag{1.3}$$

K_{vi} the utilization coefficient of the repairable ith element, where:

$$K_{vi} = \lim_{t \to \infty} A(t) = \frac{\mu_i}{\mu_i + \lambda_i} + \frac{T_i}{T_i + \theta_i} \tag{1.4}$$

K_{pi} is the downtime coefficient of the repairable ith element, where:

$$K_{pi} = 1 - K_{vi} = \frac{\lambda_i}{\mu_i + \lambda_i} + \frac{\theta_i}{T_i + \theta_i} \tag{1.5}$$

- T_i [h] mean time between two subsequent errors of the ith element,
- λ_i [h^{-1}] the mean value of the error intensity of the ith element, where,
- θ_i [h] the mean time of repair duration of the ith element,
- μ_i [h^{-1}] the mean value of the repair intensity of the ith element.

Equations (1.1) and (1.2) represent the exponential distribution of errors.

1.4 The System Reliability Characteristics

The resulting reliability characteristics of production systems formed from individual elements (machine tools, robots, manipulators, transport elements, rack stackers, etc.) must be determined from the structure of the system in which the sequence (continuity, dependence) of operations is determined by mass flow bonds. In specific cases, it is necessary to consider such facts as e.g. the possibility of repairing an individual element of the production cell [22] while the surrounding machines are running, etc.

In these cases, it is necessary to analyse in detail the individual cells of the production system using their description based on the representation of stationary Markov processes, the construction, and the solution of Kolmogorov differential equations.

For basic reliability structures, i.e. serial and parallel structure, it is possible to derive simple relations for the utilization coefficients of the system organized in these structures [12].

1.5 The Serial Structure

In this production structure, the system contains only the necessary number of elements for the realization of its function. In this case, for the error-free operation of the system, all its elements must be in the error-free state and any element functional error will cause the failure of the entire system [10].

The error rate λ(t) is given by the ratio of the probability density f(t) to the probability of fault-free operation R(t) at a given time of operation and is one of the most important characteristics in reliability theory. It will be:

$$\lambda(t) = \frac{f(t)}{R(t)} = \frac{f(t)}{1 - F(t)} = \frac{-\frac{dR(t)}{dt}}{1 - F(t)} = -\frac{\dot{R}(t)}{R(t)} \tag{1.6}$$

The error rate λ(t) at time t (since commissioning) numerically expresses the error probability per time period Δt (per unit time) in the following time $t + \Delta t$ of operation, if the monitored element or system is at time t in an error-free state. The value λ(t) of technical equipment is approximately in the range (0.01–0.0001) h^{-1}. The probability that for the monitored element (resp. System), whose intensity of failures at the time of its active operation t is $\lambda = 0.001$ will occur in the next 1 h of operation, is one-thousandth.

The typical progress of error intensity change in more complex systems is shown in Fig. 1.1. The progress of the bathtub curve may be different for devices of different structures and also depends on the operating conditions in which the device operates and the quality of maintenance, but its shape for a particular system is usually considered for normal (defined) operating conditions. In the normal life period, in systems with a complex structure and a large number of elements (components), the error rate is often close to

where the time period $\vartheta \in (0, A)$ is called "running-in time"

$\vartheta \in (A, B) \equiv t$ "Period of normal (active) life"

$\vartheta > B$ "Life expectancy"

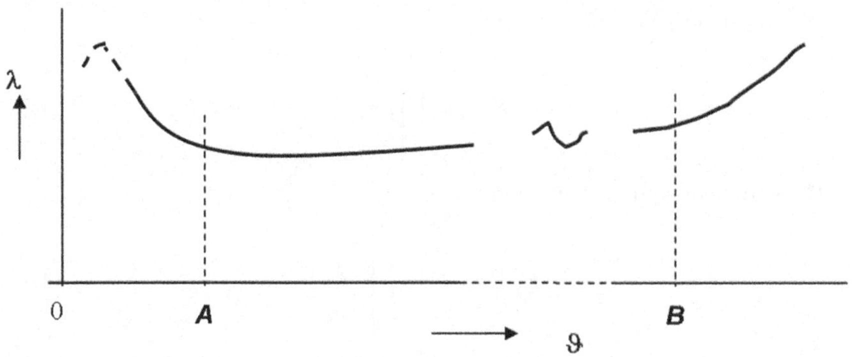

Fig. 1.1 The bathtub curve of system failures during its "life"

the constant value. At this point, it should be noted that even manufacturers of complex (especially production and transport machinery and equipment) design and manufacture products so that after the end of normal life the error rate of most functional elements increases and evokes the need to decommission the product, which sometimes has positive economic and often security significance.

The mean time between T_s failures is:

$$T_s = \frac{1}{\lambda_s} \tag{1.7}$$

The average system repair time is:

$$C\theta_s = \sum p_i\theta_i \tag{1.8}$$

where in $p_i = \frac{\lambda_i}{\lambda_s}$ is the error probability in the ith element of the system.

The error-free probability of the system $R_s(t)$ is:

$$R_S(t) = \prod_{i=1}^{n} R_i(t) \tag{1.9}$$

The system error probability $F_s(t)$ is:

$$F_S(t) = 1 - \prod_{i=1}^{n} R_i(t) \tag{1.10}$$

The system readiness (from repairable elements) is:

$$A_S(t) = \prod_{i=1}^{n} A_i(t) \tag{1.11}$$

The system utilization coefficient:

$$K_{vs} = \prod_{i=1}^{n} K_{vi} \tag{1.12}$$

The system downtime coefficient:

$$K_{ps} = 1 - \prod_{i=1}^{n} K_{vi} \tag{1.13}$$

The sensitivity coefficient of the system to change the sensitivity coefficient of the ith element:

$$\xi_{s,i} = \frac{\partial K_{vs}}{\partial K_{vi}} \tag{1.14}$$

The utilization coefficient absolute change of the system at the utilization coefficient absolute change of only i-th element ΔK_{vi} will be:

$$\Delta K_{vs,i} = \xi_{s,i} \Delta K_{vi} \tag{1.15}$$

The utilization coefficient percentual change of the system with the utilization coefficient absolute change of only the ith element will be:

$$\delta K_{vs,i}[\%] = \xi_{s,i} \frac{\Delta K_{vi}}{K_{vs}}.100 \tag{1.16}$$

The utilization coefficient percentual change of the system when the utilization coefficient of the ith element changes by p_i [%] will be:

$$\delta K_{vs,i}[\%] = \xi_{s,i} \cdot p_i K_{vi} \tag{1.17}$$

Note: The relations (1.14), (1.15), (1.16), (1.17) apply to any reliability structure must be correctly determined $\xi_{s,i}$.

1.6 The Parallel Structure

The structure is characterized by the fact that a system error occurs only when all elements fail at the same time.

The error-free operation probability of the structure consists of "n" elements is:

$$R_s(t) = 1 - \prod_{i=1}^{n} [1 - R_i(t)] \tag{1.18}$$

wherein $R_i(t)$ is the error probability of the ith element.

The error probability:

$$F_S(t) = \prod_{i=1}^{n} (1 - R_i(t)) = \prod_{i=1}^{n} F_i(t) \tag{1.19}$$

where $F_i(t)$ is the error probability of the i-th element.

The system readiness (from repairable elements):

$$A_S(t) = 1 - \prod_{i=1}^{n} (1 - A_i(t)) \tag{1.20}$$

where $A_i(t)$ is the readiness of the i-th element.

The utilization rate of the system will be:

$$K_{vs}(t) = 1 - \prod_{i=1}^{n} (1 - K_{vi}) \tag{1.21}$$

where $K_{vi}(t)$ are the individual elements utilization coefficients.

The system downtime coefficient will be:

$$K_{ps}(t) = \prod_{i=1}^{n} (1 - K_{vi}) \tag{1.22}$$

Exercise 1 Consider a device (e.g., a machine tool) in a period with a constant intensity of errors $\lambda_0 = 0.08\,\mathrm{h}^{-1}$ and an intensity of maintenance $\mu_0 = 2\,\mathrm{h}^{-1}$.

It is necessary to determine:

- mean time between errors T_0 and mean time of maintenance Θ_0 of the machine,
- utilization factor K_V and machine K_P downtime,
- probability of machine operation at time t = 8 h, if it was running at time t = 0,
- interval readiness of the machine for an operating time of 8 h (during a work shift).

Solution:

a. according to Eq. 1.7 $T_0 = \frac{1}{\lambda_0} = \frac{1}{0.08} = 12.5\,\mathrm{h}$

according to Eq. 1.8 $\Theta_0 = \frac{1}{\mu_0} = \frac{1}{2} = 0.5\,\mathrm{h}$

b. according to Eq. 1.12 $K_V = \frac{\mu_0}{\mu_0 + \lambda_0} = \frac{2}{2+0.08} = 0.96154 = 96.2\%$

according to Eq. 1.13 $K_P = 1 - K_V = 0.03846 = 3.8\%$

c. $A(t) = \frac{\mu_0}{\mu_0+\lambda_0} + \frac{\lambda_0}{\mu_0+\lambda_0}.e^{-(\mu_0+\lambda_0)\cdot t}$

$$A(8) = \frac{2}{2+0.08} + \frac{0.08}{2+0.08}.e^{-(2+0.08).8} = 0.962$$

d. $E(t) = \frac{\mu_0}{\mu_0+\lambda_0} + \frac{\lambda_0}{T(\mu_0+\lambda_0)^2}.\left(1 - e^{-(\mu_0+\lambda_0)T}\right)$

$$E(8) = 0.96154 + \frac{0.08}{8 \cdot (2.08)^2}(1 - e^{-8\cdot 2.08}) = 0.964$$

The above example shows the probability that with a sufficiently long period of use (after running-in) the machine will be 96.2% of the time running and 3.8% of the time the errors will be rectified (performed repair). The probability that the machine will operate in the time since start-up is 0.962 and the probability that it will operate without failure for 8 h (during the working change) is 0.964.

1.7 Serial and Parallel Structure

(a) **The serial structure (basic assembly)**

A system with a series structure is formed by elements arranged one behind the other, i.e. the output of the previous element is the input for the next element, etc. The structure is shown in Fig. 1.2 and the elements of the structure are indicated by the capital letter E.

The transition between adjacent process states (or system elements) expresses the intensity λ.

An error in a system with a series structure occurs whenever an error occurs in any of its components. If we assume the independence of the occurrence of failures for individual elements, then the probability of fault-free operation of the system at the time of its use will be:

$$R_s(t) = \prod_{1}^{n} R_k(t) \tag{1.23}$$

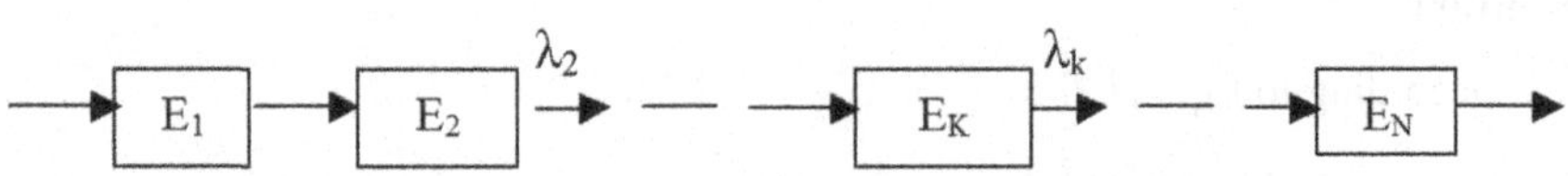

Fig. 1.2 The system serial structure

where:

$$\lambda_s = \sum_{1}^{n} \lambda_k \tag{1.24}$$

And the error probability:

$$\mathrm{R_s(t) = 1 - R_s(t)} \tag{1.25}$$

Equation (1.24) shows that the intensity of system errors, which indicates the number of errors per unit time, increases with an escalating number of elements in the system. The intensity of system faults is given by the sum of the intensity of errors in the system.

In technical practice, the system error is usually considered any state in which the system is not able to fully perform the required operation. If the readiness of any element of the serial structure is considered to be an independent phenomenon (idealization) then the resulting readiness of the system will be given by the product of individual phenomena, so it will be:

$$A_s(t) = \overset{n}{\underset{1}{\pi}} A_k(t) = \overset{n}{\underset{1}{\pi}} \left\{ \frac{\mu_k}{\mu_k + \lambda_k} + \frac{\lambda_k}{\mu_k + \lambda_k}^{[-(\mu_k + \lambda_k).t]} \right\} \tag{1.26}$$

Different solution for the system readiness calculation is used because determining the system readiness by analysing its functional structure is sometimes too demanding (especially for systems with many elements). We will consider the system as a whole with mean values of error and repairability characteristics, which we determine as follows.

The mean time of error-free operation of the system $\mathrm{T_s}$ is determined from the results of monitoring the operation of the system as a ratio of the total actual time of error-free operation $T = \sum_1^k T_k$ to the number of K errors that occurred in the monitored time interval, i.e.:

$$T_s = \frac{1}{k} \sum_{1}^{k} T_k$$

and *mean value of the error intensity* (with exponential distribution):

$$\lambda_s = \frac{1}{T_s} = \frac{k}{\sum_1^k T_k} \tag{1.27}$$

If we assume that after each failure, the system will be repaired immediately, then the mean time of repairs from monitoring in the same time period can be determined

$$\Theta_s = \frac{1}{k}\sum_{1}^{k}\Theta_k$$

as *the mean value of the intensity of repairs:*

$$\mu_s = \frac{1}{\Theta_s} = \frac{k}{\sum_{1}^{k}\Theta_k} \tag{1.28}$$

The relationship for determining (predicting) the operational readiness of the system in the period of its subsequent operation, assuming that at time t = 0 it was functional, then it will be:

$$A_s(t) = \frac{\mu_s}{\mu_s + \lambda_s} + \frac{\lambda_s}{\mu_s + \lambda_0}^{[-(\mu_s+\lambda_s).t]} \tag{1.29}$$

We also use the same procedure to determine other readiness characteristics, such as:

the system interval readiness over time $\tau = T$

$$E_s(t) = \frac{\mu_s}{\mu_s + \lambda_s} + \frac{\lambda_s}{T(\mu_s + \lambda_0)^2}\{^{1-[-(\mu_s+\lambda_s)T]}\} \tag{1.30}$$

the system utilization factor:

$$K_{Us} = \frac{\mu_s}{\mu_s + \lambda_s} = \frac{T_s}{T_s + \Theta_s} \tag{1.31}$$

the system downtime factor:

$$K_{ps} = \frac{\lambda_s}{\mu_s + \lambda_s} = \frac{\Theta_s}{\Theta_s + \Theta_s} \tag{1.32}$$

To assess the characteristics of the system with the basic assembly (serial structure), it is important to know that the error rate, but also the readiness of the system is significantly negatively affected by the most faulty and longest repaired (most failed) element of this structure (applies in general). Therefore, from the point of view of the readiness of a system with a serial structure, it is suitable and optimal to use elements with approximately the same error and repair intensities. At the same time, this results in the knowledge that in a structure with unsuccessful functional elements (have no better), it is not appropriate to use elements with extremely good parameters (expensive) at the same time. Using derived relationships, for specific systems, it is possible to analyse in detail the impact of changes in the intensity of errors and repairs on the utilization coefficients, for instance, downtime and their consequences.

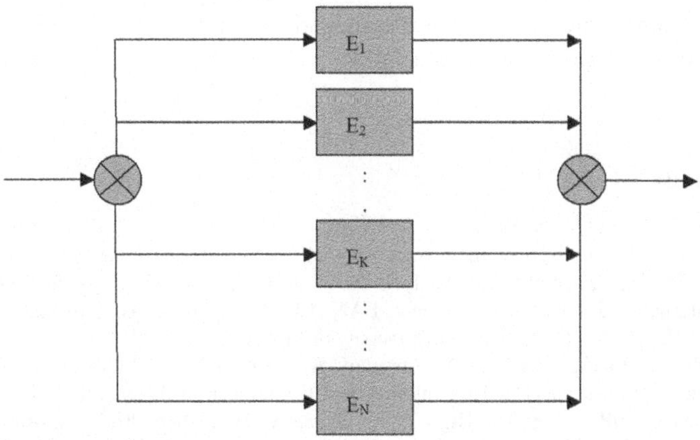

Fig. 1.3 The system parallel structure

(b) **Parallel structure (backup assembly)**

To increase the probability to perform the system operation, it is required that with initiated system input, the desired output will be provided even if only one path from input to output is functional. We then create the system from parallel branches. Such a parallel structure can realize the given requirement. A diagram of the structure is shown in Fig. 1.3.

The probability that the system will be functional (input → output) will be given by the probability that the error of all its elements will not occur at the same time. Since the errors of the system individual elements are independent phenomena, error and error-free operation are opposite phenomena, the probability of error-free system operation (opposite to failure) will be consistent with the application of de Morgan's rule, given by the product of phenomena opposite to the probability of error of any element [9]:

$$1 - R_S(t) = \prod_{1}^{N} [1 - R_K(t)] \tag{1.32}$$

from where for *the probability of error-free operation of the parallel system $R_S(t)$* follows:

$$R_S(t) = 1 - \prod_{1}^{N} [1 - R_K(t)] \tag{1.33}$$

Because the expression after the product symbol on the right side of Eq. (1.33) is always a number less than 1, with increasing numbers in the structure, $R_S(t)$ converges rapidly to 1, which means that the probability of error-free operation of the system is close to certainty. The readiness of the structure is not disturbed by the repair of one or more elements, and therefore it is theoretically possible to achieve 100% readiness of the system by using a backup assembly.

References

1. Abramov, I., Božek, P., Abramov, A., Sosnovich, E., Nikitin, Y.: Diagnostics brushless DC motors. In: Experimental Stress Analysis 2017 [electronic source]: Conference Proceedings. 55th International Scientific Conference. EAN 2017, Nový Smokovec, Slovakia, 30 May to 1 June 2017, pp. 156–164. Technical University of Košice, Košice (2017)
2. Abramov, I., Božek, P., Nikitin, Y., Abramov, A., Sosnovich, E., Stollmann, V.: Diagnostics of electrical drives. In: The 18th International Conference on Electrical Drives and Power Electronics. EDPE 2015. The High Tatras, Slovakia, 21–23 Sept 2015, pp. 364–367. IEEE [b.m.] (2015)
3. Abramov, I., Nikitin, Y., Zorina, O., Božek, P., Stepanov, P., Štollmann, V.: Monitoring of technical condition of motors and bearings of woodworking equipment. Acta Facultatis Xylologiae Zvolen **56**(2), 97–104 (2014)
4. Basseville, M., Nikiforov, I.V.: Detection of Abrupt Changes: Theory and Application. Prentice Hall Information and System Sciences Series, 447 p. Prentice Hall, Englewood Cliffs, NJ (1993)
5. Cech, M., Konigsmarkova, J., Goubej, M., Oomen, T., Visioli, A.: Essential challenges in motion control education. In: 12th IFAC Symposium on Advances in Control Education (ACE), Philadelphia, PA, 07–09 July 2019, vol. 52, no. 9, pp. 200–205 (2019). ISSN: 2405-8963
6. Chang Liang, X.: Permanent Magnet Brushless DC Motor Drives and Controls, 295 p. Science Press, Wiley, Singapore (2012)
7. Frank, P.M.: Fault diagnosis in dynamic systems using analytical and knowledge-based redundancy: a survey and some new results. Automatica **26**(3), 459–474 (1990)
8. Gizelska, M., Kozanecka, D., Kozanecki, Z.: Diagnostics of the mechatronic rotating system. Key Eng. Mater. **588**, 101–108 (2014). ISSN: 1013-9826
9. Gulyashinov, A.N., Tenenev, V.A., Yakimovich, B.A.: Teoriya prinyatiya reshenij v slozhnyh sociotekhnicheskih sistemah (Decision Theory in Complex Socio-technical Systems), 280 p. Izhevsk, IzhGTU (2005)
10. Isermann, R.: Fault-Diagnosis Systems: An Introduction from Fault Detection to Fault Tolerance, 475 p. Springer, Berlin (2006)
11. Isermann, R.: Identification of Dynamic Systems. An Introduction with Applications, 705 p. Springer, Berlin, Heidelberg (2011)
12. Jones, H.L.: Failure detection in linear systems. Ph.D. thesis, MIT, MA, USA (1973), 459 p
13. Lacko, B., Beneš, P., Maixner, L., Šmejkal, L.: The Automation and the Automation Technology. Part 1: The Automation System Concept. Computer Press, Praha (2000). ISBN: 80-7226-246-7
14. Maixner, L.: The Automatic Production System Design. SNTL, Prague (1980)
15. Mozgalevskij, A.V., Kalyavin, V.P., Kostandi, G.G.: Diagnostirovanie elektronnyh system (Diagnostics of Electronic Systems), 224 p. Leningrad, Sudostroenie (1984)
16. Nikitin, Y., Turygin, Yu., Sosnovich, E., Božek, P.: Trends in control of NC machines. In: Procedia Engineering [Electronic Source]: International Conference on Manufacturing Engineering and Materials, ICMEM 2016, Nový Smokovec, Slovakia, 6–10 June 2016, vol. 149, pp. 352–358 (2016)
17. Sinopal'nikov, V.A.: Nadezhnost' i diagnostika tekhnologicheskih system (Reliability and Diagnostics of Technological Systems), 343 p. Vysshaya shkola, Moscow (2005)
18. Skhirtladze, A.G.: Nadezhnost' i diagnostika tekhnologicheskih sistem (Reliability and Diagnostics of Technological Systems), p. 518. Novoe znanie, Moscow (2008)
19. Soffker, D., Wolters, K., Ozbek, M., Dettmann, K.U.: Feature-based diagnosis and prognosis for an integrated diagnostic approach. In: Chang, F.K. (ed.) 6th International Workshop on

Structural Health Monitoring, Stanford Univ., Stanford, USA, 11–13 Sept 2007, pp. 754-761. Stanford Univ. (2007). ISBN: 978-1-932078-71-8
20. Timofeev, A.V.: Physical diagnostics and fault relevant feedback control. In: International Conference on Physics and Control (PHYSCON 2003), St. Petersburg, Russia, 20–22 Aug 2003, pp. 253–258 (2003). ISBN: 0-7803-7939-X
21. Turygin, Y., Božek, P., Abramov, I., Nikitin, Y.: Reliability determination and diagnostics of a mechatronic system. Adv. Sci. Technol. Res. J. **12**(2), 274–290 (2018)
22. Turygin, Y., Božek, P., Nikitin, Y., Sosnovich, E., Abramov, A.: Enhancing the reliability of mobile robots control process via reverse validation. Int. J. Adv. Rob. Syst. **13**(6), 1–8 (2016)
23. Vekteris, V., Cereska, A.: Diagnostical measurements of elements of mechatronical system. In: Kyttner, R. (ed.) Proceedings of the 5th International Conference of the Danube-Adria-Association-for-Automation-Manufacturing, Tallinn, Estonia, 20–22 Apr 2006, p. 183 (2006). ISBN: 9985-894-92-8
24. Vrban, A.: The Management System Reliability. Course Literature. STU, Bratislava (2007). ISBN 978-80-8096-010-0
25. Vrban, A.: The Production Machines and Systems Reliability. Course Literature. SVŠT, Bratislava (1983)
26. Vrban, A.: The Technical Systems Diagnostics. Course Literature. SVŠT, Bratislava (1985)
27. Vrban, A., Božek, P.: The multitasking drilling machine analysis and reliability forecast. HZ No. 48/88, Trnava (1988)
28. Zhirabok, A.N., Shumskii, A.E., Solyanik, S.P., Suvorov, AYu.: Design of nonlinear robust diagnostic observers. Autom. Remote Control **78**(9), 1572–1584 (2017)

Chapter 2
Methods, Models, Algorithms for Diagnostics of Mechatronic Systems

2.1 Methods of Mechatronic Systems Diagnostics

Methods of mechatronic systems diagnostics are classified according to the following features [1, 3]:

- Information content rate;
- Types of diagnostic information;
- Technical equipment usage rate;
- Operation phase;
- Depth of diagnosis.

The following methods of mechatronic systems diagnostics are characterized by the rate of information content [10]:

- *the time interval method* used for downtime analysis, determination of reliability indicators [5], control of the operation of the control system, acquisition of cyclograms;
- *the method of reference modules* based on a comparison of experimental data or calculated values and quality indicators;
- *the standard dependency method* based on comparison of measured diagnostic parameters with reference diagnostic parameters;
- *the spectral method* based on measuring the components of complex vibrational or acoustic signals;
- *the correlation method* used to detect deviations in like the relationship between diagnostic parameters (cross-correlation) or changes in diagnostic parameters over time (autocorrelation).

According to the source of information for diagnostics, a distinction is made between test and functional diagnostics. During test diagnostics, the test effects of diagnostic tools are applied to the object. During functional diagnostics, only work exposures reach the object. Diagnostic methods are classified according to the

P. Božek et al., *Diagnostics of Mechatronic Systems*, Studies in Systems, Decision and Control 345, https://doi.org/10.1007/978-3-030-67055-9_2

type of physical processes occurring in the object: mechanical, electrical, vibration, ultrasonic, shock, thermal, magnetic, photometric, etc.

The mechanical method consists of measuring the geometric dimensions of parts of mechanical objects. This method is used to determine the wear of the mechanical parts.

The electrical method includes measuring electrical quantities. For example, the electrical diagnostic method is based on measuring electrical diagnostic parameters: current, voltage, resistance, power.

The vibration method involves measuring the parameters of vibration processes - vibration shift, vibration speed, vibration acceleration.

The ultra-sonic method contains measuring directional ultrasonic oscillations reflected from the interface of two media. This method allows you to detect internal defects of large objects with the ability to determine the position and size of defects.

The method of shock pulses consists of the measurement of mechanical shock waves generated during the collision of solids. This method is used to determine the condition of rolling bearings and lubrication.

The thermal method consists of measuring the temperature of objects.

The magnetic method includes measuring the magnetic dispersion fields generated above the error and determining the magnetic properties of the diagnosed objects. This method is used to measure the thickness, to check the structure and mechanical properties of parts. Depending on the method of measuring magnetic fields, the dispersion is divided into magnetic particles, ferrozine resistors, magnetic resistors, inductions.

The photometric method is the measurement of light. It is used to measure linear and angular play and play at the interface, oil pollution.

Promising areas of the development of the diagnostic methods are currently artificial intelligence methods based on fuzzy logic, expert systems, and neural networks. Fuzzy logic methods can greatly simplify the description of the model of diagnosed objects and are also easier to implement hardware. Expert systems allow you to make decisions about the state of an object (if it is difficult to assess the state or eliminate the problem with the object). Neural networks are used to identify objects, recognize and predict the state of the mechatronic system. The advantages of a classifier based on neural networks over traditional estimation methods are in such factors:

- noise independence,
- self-study,
- possibility of parallel processing.

An important step in any method of diagnosis is the creation of a mathematical model that provides adequate information about the functioning of the mechatronic system.

2.2 Diagnostic Models of Mechatronic Systems

The success of the mechatronic system diagnostics, to a large extent, depends on the correct choice of node diagnostic models and diagnostic functions. Based on the properties of the diagnostic object and the conditions of its operation, diagnostic models are being developed, the effectiveness of which largely depends on the degree of adaptability of the object design for technical diagnostics, as well as the methods and means of technical diagnostics used.

Diagnostic models are called models for the theoretical analysis of objects. One of the classifications of diagnostic models is given in the work of A. V. Mozgalevsky [2].

Diagnostic models are divided into three groups:

1. continuous;
2. discrete;
3. special.

Diagnostic models are divided into three groups:

1. algebraic equations;
2. differential equations.

Discrete diagnostic models are divided into two subgroups:

1. definitive differential equations;
2. definitive automata.

Special diagnostic models are divided into three subgroups:

1. information;
2. functional;
3. static and dynamic characteristics.

Analytical diagnostic models are widely used to describe all nodes, for example mechanical, electrical, electromechanical, or pneumohydraulic. The main problem in the development of analytical diagnostic models is the determination of the difference between the actual and the reference value of diagnostic parameters. Most determinants of differences are based on models of a linear system. For nonlinear systems, the main approach is their linearization. However, for systems with a high degree of nonlinearity and many nonlinear operations, this linearization does not provide satisfactory results.

The solution to this problem is to use a large number of linear systems, which is not very practical when creating models that work in real time. Most known linearization is only applicable to a limited class of nonlinearities. Besides, the modelling process is very complicated, and the accuracy of the results obtained is difficult to verify. Neural networks are suitable for solving this problem.

If it is not possible to use analytical methods due to lack of information or the structure of the object itself, it is recommended to use topological models where the parameters are presented as a graph or matrix with causal links.

Analytical diagnostic models make it possible to apply appropriate optimization methods and obtain ratios that characterize an object when its state changes. Graphical diagnostic models have greater visibility and can serve both directly and to illustrate analytical methods. Graphoanalytic diagnostic models are various combinations of graphical and analytical models.

Consideration of object characteristics makes it possible to reduce the number of suitable diagnostic models for mechatronic systems. For example, for machine tools, models describing individual objects are not considered. These models include truth functions, logical networks, graphs. The use of graph-topological, functional models, structural and equivalent schemes requires detailed knowledge of the functional structure and cause-effect relationships in the diagnosed object. Therefore, the use of such models is difficult.

It most fully corresponds to the characteristics of the objects of the diagnosed model of recognition theory. The creation of models takes place according to the training sequence and does not require knowledge of exact analytical dependencies between input, internal, and output parameters.

There are two approaches to creating recognition models:

- probabilistic and
- deterministic.

The probabilistic approach requires knowledge of the multidimensional densities of the conditional probability distribution for each class. Under these conditions, statistical solution methods are applicable that test static hypotheses about class descriptions to states with minimal risk for an erroneous solution. Depending on the choice of risk criterion, specific methods of statistical solutions are distinguished: the minimum risk, minimum number of wrong decisions, the maximum probability of minimax method, Neumann method, Pearson, and others.

If the analytical form of the conditional probability density formulas is not known and its exact definition is not assumed, then it is assumed that the distribution is uniform within a certain region. This approach is adopted in Bayesian and sequence analysis methods. The sequence analysis method is used in dichotomy and makes it possible to reduce the number of elementary controls.

Models of discriminant analysis are based on the concepts of discriminant and separation functions. The state class is selected at the maximum value of the resolution function.

Cluster analysis models are based on the fact that class areas in the diagnostic space form natural clusters. The ability to set different metrics and similarity assessment methods remotely creates a wide range of diagnostic methods.

Synergetic methods make it possible to describe processes in complex systems of various nature using some universal representations and models. For example, to estimate the technical condition of the device and reliably identify such errors as imbalance, shaft deflection, loss of support stiffness, it is possible to use the method of construction of phase portraits based on the theory of deterministic chaos.

Models and diagnostic methods are very diverse. At the same time, not all are acceptable for building systems for the diagnosis of mechatronic systems.

2.3 Algorithms for the Mechatronic Systems Diagnostics

An overview of existing diagnostic algorithms and software shows the absence of universal algorithms for diagnosing mechatronic systems. Neural networks are a promising mathematical apparatus for creating systems for automatic diagnostics of mechatronic objects. Neural networks have the following advantages:

- fast learning algorithms,
- ability to work in the presence of significant interference,
- ability to work with various information,
- ability to solve several problems at the same time (parallelism of information processing),
- reliable operation.

The main advantage of the neural network approach is the ability to identify patterns in the data, their generalization, i.e. gain knowledge from data. Statistical methods for detecting the state of the mechatronic system are used in cases where a mathematical model of physical processes is not known or cannot be obtained. For example, when physical processes are not reliably described by known equations of mathematical physics or size exceeds the capacity of the existing model of computing. At the same time, statistical methods of state recognition require a considerable amount of a priori data, i.e. data obtained as a result of experiments. Therefore, for expensive experiments, statistical methods of state recognition may be unacceptable.

Diagnosis of system errors using deterministic state recognition methods is effective in the presence of a mathematical model of its operation. In most cases, these models can only be analysed by numerical methods, which limit their use in real time in troubleshooting and managing the technical system. Almost all real processes of operation of technical systems have nonlinear behaviour. In these cases, experts are usually used, i.e. the person intervenes in the process of diagnosing and managing the technical system. If deterministic knowledge is not available or mathematical modelling is expensive or does not provide the required accuracy, other methods may be used. Such methods are modelling of operator knowledge using heuristic knowledge and logical inference strategies, such as in expert systems based on fuzzy logic with their implementation based on hardware or software-algorithmic emulation neural networks.

An important feature of neural networks is that they study the dynamics of the system in the training process, which consists of several training cycles, while the training data come either from the previous cycle or consist of real signals. After each cycle, the neural network learns more and more about the dynamics of the object. One of the most important properties of neural networks is their ability to automatically study the behavioural dynamics of nonlinear systems when the architecture of a neural network contains at least three layers. It is possible to use neural networks in problems of predicting the remaining lifetime of MS.

According to the use of the mathematical apparatus, diagnostic algorithms are divided into 3 groups:

- algorithms in which the measured diagnostic parameter is compared with a threshold value;
- algorithms that use the processing of the measured diagnostic parameter using Fourier transforms, Haart and others (kepstr, wavelet);
- algorithms where soft computing methods are used based on artificial neural networks, fuzzy set theory, genetic algorithms.

According to the mode of operation, the real-time diagnostic algorithms are divided into two groups:

1. on-line;
2. off-line.

The diagnostic algorithm determines the composition and diagnostic procedure of the mechatronic system.

The following types of diagnosis are commonly used:

- elementary, consisting of diagnostics of each element separately;
- modular, consisting of diagnostics of individual mechatronic modules of the system, each of which consists of several elements;
- group, including diagnostics of a group of elements that are interconnected but do not form a mechatronic model.

To create a rational diagnostic algorithm, you must:

1. represent a mechatronic system in the form of a system that reflects the individual functional elements and the links between them;
2. identify a list of all possible errors and provide a formal description of the operation of the system;
3. compile a mathematical description of the faulty system;
4. develop a rational diagnostic algorithm.

The compilation of diagnostic algorithms is greatly simplified if the specification of the defect list indicates the place of design where a typical error is possible [6, 12]. It is often necessary to find a place where the defect formed and determine its cause.

The diagnostic algorithm often includes the following sequence:

- first, the main characteristics of the mechatronic system are measured and it is determined whether it operates in a given mode or whether there are excessive deviations.
- It also determines the cause of these deviations and uses special tests (checks) to determine the error.

Thus, in the first stage, a functional diagnostic is performed and then the diagnostic is tested for an error.

When diagnosing electrical and electronic components, it is necessary to determine the error at the level of a separate element (resistor, diode, transistor, etc.). In this case, a high-resolution test will be required to assess system performance [8].

Because mechatronic systems are built according to a modular principle, module diagnostics is sometimes sufficient to diagnose them. A remedy in this case is achieved by replacing the faulty module with a service module.

When monitoring the performance of the diagnostic system, it must objectively determine whether the component or module is intact or defective [7]. Checking the correctness of operation consists in determining how the module works in the current time and whether its parameters correspond to good technical condition [9].

At present, artificial intelligence methods such as expert systems, artificial neural networks, fuzzy logic methods, and genetic algorithms are used as mathematical apparatus for diagnostics. Neural networks represent a nonlinear model without knowledge of its structure and produce results in a short period of time. The inputs of neural networks are current, voltage, power, temperature, vibrations, spatial position accuracy, and rigidity, motion parameters, power parameters, time intervals [4]. The main problem when working with a neural network is the selection of the best input functions and parameters of the neural network that make it compact, and the classification of defects is accurate.

The inference engine of the expert system classifies the state of the object using a database that contains a history of the state of the device, which would describe trends in typical types of faults. Knowledge of the tendency to change the diagnostic parameter of any type of fault is crucial because it allows the expert system to conclude whether the error is accepted or ignored based on a threshold value [11].

The key decision in the diagnosis of the current situation is the choice of an effective classification system. They can be divided into two main groups:

1. knowledge-based models,
2. data-based models.

The diagnostic tool with the ability to dynamically acquire knowledge is required. The presence of error examples of in not demanded when the implementation of correct diagnostics is adapted and the tool is applicable to different typologies of equipment, at least for machines of the same series.

It is often difficult to collect data that reflects the entire "error domain", while it is easier to identify an area of "operational value". Modern approaches are based on neural networks, trained on data taken from a normally functioning installation, and are able to detect errors based on data that is outside the area defined during the training.

Neural networks make it possible to effectively determine the causes and types of damage to mechatronic systems, work with noisy data, eliminate the need to use intermediate electronic filters before interference or filtering using mathematical methods, and also adapt to a specific instance.

Intelligent diagnostic systems are built as self-learning systems with flexible decision-making, such as knowledge-based systems that create new knowledge in the process of functioning. The tasks of the intelligent diagnostic system include assessment of the technical condition, analysis of the functioning environment, etc. The class of intelligent diagnostic systems complies with the following five principles:

1. interaction of the diagnostic system with the real outside world using information communication channels [2]. Intelligent diagnostic systems derive knowledge from it and influence it. Implementing this principle allows you to organize a communication channel to gain knowledge and organize appropriate behaviour;
2. basic exposure of the system in order to increase intelligence and improve own behaviour (exposure of the system is ensured by the presence of self-adaptation, self-organization, and self-study). The knowledge system of an intelligent diagnostic system consists of two parts: incoming knowledge and proven knowledge. This principle makes it possible to organize the supplementation and acquisition of knowledge;
3. The mechanism for the availability of mechanisms to predict changes in the operating environment and the system's own behavior in a dynamically changing external world. In line with this principle, an intelligent diagnostic system is not fully intelligent if it does not have the ability to predict changes in the outside world and its own behaviour;
4. the system has a building structure that corresponds to the principle of IPDI (Increasing Precision with Decreasing Intelligence): the accuracy of control is higher the less intelligence of the system. This is a way to build complex intelligent diagnostic systems in case the inaccuracy of knowledge about the model of the control object or its behaviour can be compensated by increasing the intelligence of the created system;
5. maintaining operation in the event of interruption or loss of control actions from higher levels of the hierarchy.

For example, an intelligent system for CNC machine diagnostics is based on an information system for analysing the operation of process equipment for shaping quality parameters using artificial neural networks, which are the knowledge base of the system and optimization of control activities (cutting mode parameters) using a genetic algorithm. The system consists of the following function blocks:

- information system for the analysis of the operation of technological equipment on the qualitative parameters of shaping using artificial neural networks. It is implemented as a program for determining quality parameters;
- control optimization systems based on a genetic algorithm;
- expert system.

Genetic algorithms do not guarantee the detection of a global solution in the shortest possible time. Genetic algorithms do not guarantee that a global solution can be found, but they are good for finding a "good enough" solution to a problem "relatively quickly".

These phases of solving problems of mechatronic systems diagnostics using a neural network are emphasized:

- data collection for training;
- preparation and standardization of data;
- Network topology selection;

Table 2.1 Usage of neural networks for machine nodes diagnostics (table cells indicate the number of publications that describe the successful solution of the diagnostic problem using this network)

The neural network type	Diagnostic objects						
	Pumps	Bearings	Gears	Dosing boxes	Rotating systems	Vents	Motors
BPFF	1	1	–	–	–	–	–
FFNN	1	–	–	–	–	–	1
RNN	1	–	–	–	–	–	1
RBF	–	1	–	–	1	–	–
BP	2	1	1	1	1	–	–
MLP	–	2	2	1	2	1	1
SOM Kohonen	–	–	–	–	–	–	1
LVQ	–	1	1	–	1	–	–

1. experimental selection of network characteristics;
2. experimental selection of learning parameters;
3. proper network training;
4. checking the adequacy of the training;
5. parameter setting, final training;
6. Ization Verbalization of the network for further use.

An analysis of the publications [3–10] showed that there are many successful applications of neural networks for machine node diagnostics. The results of this analysis are shown in Table 2.1:

BPFF (Back Propagation for Feed-Forward Networks)—forward propagation networks with an error propagation algorithm;

FFNN (Feed Forward Neural Networks)—direct distribution neural networks;

RNN (Recurrent Neural Networks);

RBF (Radial Basis Function Network);

BP (Back Propagation)—networks with reverse error propagation algorithm;

MLP (Multilayer Perceptron);

SOM (Self-Organising Map);

LVQ (Learning Vector Quantization).

References

1. Abramov, I., Nikitin, Yu., Abramov, A., Sosnovich, E., Božek, P.: Control and diagnostic model of brushless DC motor. J. Electr. Eng. **65**(5), 277–282 (2014)
2. Auslander, D.M., Huang, A.N., Lemkin, M.: A design and implementation methodology for real-time control of mechanical systems. Mechatronics **5**(7), 811–832 (1995). ISSN: 0957-4158
3. Blanke, M., Kinnaert, M., Lunze, J., Staroswiecki, M.: Diagnosis and Fault Tolerant Control, p. 571. Springer, Berlin (2003)

4. Božek, P., Shchenyatsky, A., Turygin, Y., Nikitin, Y., Karavaev, Y.: Reverse validation of a programmed robot trajectory based on INS. In: Elektro 2018 [electronic source]: 12th International Conference, Mikulov, Czech Republic, 21–23 May 2018: Conference Proceedings. IEEE, Piscataway, 4 p. (2018)
5. Dalla Vedova, M.D.L.: Diagnostic/prognostics strategies applied to physical dynamic systems: critical analysis of several model-based fault identification methods. In: Bardis, N. (ed.) 2nd International Conference on Mathematical Methods & Computational Techniques in Science & Engineering, Cambridge, GB, 16–18 Feb 2018. AIP Conference Proceedings, vol. 1982 (2018). ISSN 0094-243X. ISBN 978-0-7354-1698-7
6. Dietel, F., Schulze, R., Richter, H., Jakel, J.: Fault detection in rotating machinery using spectral modelling. In: MECATRONICS REM 2012; 9th France-Japan and 7th Europe-Asia Congress on Mechatronics (MECHATRONICS)/13th International Workshop on Research and Education in Mechatronics (REM), 21–23 Nov 2012, Supmeca, Paris, 2012, pp. 353-357, ISBN 978-1-4673-4771-6
7. Gajek, A., Strzepek, P., Dobaj, K. Algorithms for diagnostics of the hydraulic pressure modulators of ABS/ESP systems in stand conditions. In: Kalaczynski, T., Zoltowski, M. (eds.) 17th International Conference on Diagnostics of Machines and Vehicles, Bydgoszcz, Poland, 25–26 Sept 2018. MATEC Web of Conferences, vol. 182, (2018). ISSN: 2261-236X
8. Mayer, D., Atzrodt, H., Herold, S., Thomaier, M.: An approach for the model based monitoring of piezoelectric actuators. In: Computers & Structures II ECCOMAS—Thematic Conference on Smart Structures and Materials, 18–21 July 2005. Tech Univ Lisbon, Inst Superior Tecn, Lisbon, Portugal, vol. 86(3–5), pp. 314–321 (2008). ISSN: 0045-7949
9. Medjaher, K., Zerhouni, N.: Hybrid prognostic method applied to mechatronic systems. Int. J. Adv. Manuf. Technol. **69**(1–4), 823–834 (2013). ISSN: 0268-3768
10. Nikitin, Y., Bozek, P., Peterka, J. Logical-linguistic model of diagnostics of electric drives with sensors support. Sensors **20**(16) (2020). ISSN: 1424-8220
11. Vrban, A., Božek, P., Janíček, D.: The machines reliability determining methods research. Research task No. F 2106-745, Trnava (1988)
12. Zhirabok, A., Shumsky, A.: Fault diagnosis in nonlinear mechatronic systems via linear methods. In: Proceedings 42nd Annual Conference of the IEEE-Industrial-Electronics-Society (IECON), Florence, Italy, 24–27 Oct 2016, pp. 406–411. IEEE Industrial Electronics Society (2016). ISBN: 978-1-5090-3474-1

Chapter 3
Model Systems for Diagnosticing of Mechatronic Objects

3.1 Models of Information Processes for Diagnostics of Mechatronic Systems

The Mechatronic Object Diagnostic System is a software and hardware complex that consists of multiple sensors, analog-to-digital converters, a computing device that processes information and decides on the technical condition. Accordingly, information process models and hardware models for systems diagnosing mechatronic objects are considered. Analysis of existing diagnostic models of mechatronic systems (MS) makes it possible to conclude that there are no formulas for models, information processing algorithms, and decision making.

There are several information processes in the diagnostic system [1, 2]. The first information process determines how the diagnosis is organized. Determines the intervals, sequence of diagnostic modules, nodes, and MS elements. The second information process is used to decide on the technical condition of modules, nodes, and MS elements based on artificial intelligence methods.

Information process models for determining the method of diagnostics
Three models of MS diagnostic methods are proposed: parallel, sequential, and combined.

- *Model of parallel MS diagnostics.* All nodes and MS elements are diagnosed simultaneously. The technical condition monitoring is continuous. This method requires maximum cost—each node and feature has its own diagnostic microsystem. A parallel way of organizing the diagnosis process is recommended when there is a threat to human health and life. Required is high reliability because there are nodes and elements with high rates of speed of degradation processes.
- *Model of sequential MS diagnostics.* Model of sequential MS diagnostics. All nodes and MS elements are diagnosed, one by one. Regular monitoring of the technical condition is implemented. Nodes and elements have integrated sensors to measure diagnostic parameters. There is one regulator for information processing

P. Božek et al., *Diagnostics of Mechatronic Systems*, Studies in Systems, Decision and Control 345, https://doi.org/10.1007/978-3-030-67055-9_3

that processes information at certain diagnostic intervals. This method requires minimal costs. However, it is necessary to calculate the intervals of diagnosis of nodes, elements, and to determine the sequence of their diagnosis.

- *Model of combined MS diagnostics.* The most critical nodes and elements of the mechatronic system are diagnosed simultaneously and the rest—at certain intervals of diagnosis. The criterion of minimum economic losses during MS operation must be used as a target function in the selection of the method of organizing the diagnostic process. Economic losses consist of equipment downtime, STOP in product manufacturing, and diagnostic system costs [21].

Model of information process of decision-making on the technical state of MS
In order to decide on the technical state of MS, it is necessary to analyze the information provided by sensors. The following diagnostic parameters are used: electrical current, voltage, power and temperature fields, vibroacoustic parameters, spatial position accuracy, firmness, performance parameters, time periods (intervals) [9, 13, 24]. Diagnostic parameters have distinct physical characteristics; therefore, only mathematical apparatus is needed for their analysis. Neural networks, fuzzy logic, and genetic algorithm are suitable as such devices [15, 22].

The neural network that consists of modules corresponding to each MS node can be used to diagnose MS. Each module processes received information according to its diagnostic parameters. The neural network consists of layers:

- input layer, which receives sensor information and operating mode information;
- hidden layer, in which are the received data processed, the scales are assigned in the learning process;
- output layer.

In the input layer, the number of neurons and the diagnostic parameters are determined in each module. In the hidden layer, the number of neurons is equal to the number of neurons in the previous network. Every module that has its own input layer has no reference to other modules. The output layer is assigned to each module separately. The output data is a matrix whose rows correspond to the state of a diagnosed node and the columns correspond to defects in that node.

The most suitable is the development of diagnostic systems in the design phase of mechatronic modules [10]. The mechatronic module has the following advantages:

- using the one type of unified nodes;
- shortening the repair time due to node replacement;
- extension of system functions by adding mechatronic modules;
- simplified services.

So far, anti-friction bearings in an MS with integrated microprocessors have been developed and used, which measure the angular position and the rotational speed. It is possible to create intelligent electric propulsion that, in addition to MS, will have integrated the diagnostic system for detecting a defect of windings, rotor, mechanical gears, powerful semiconducting devices [3]. The main methods used for fault detection of an electrical engine in MS are vibration diagnostics, analysis of engine´s actual electrical current, measurement of the electromagnetic field of the engine with measuring coils, chemical analysis, temperature, torque and power measurement, infrared measurement, acoustic noise measurement, radiofrequency measurement, measurement of partial discharges. The most common methods are vibration diagnostics, analysis of the engine's actual electrical current, and temperature analysis due to simplicity, high accuracy, and reliability [14].

In many cases, the vibration methods are effective in detecting defects of electrical engines. However, vibration sensors, such as accelerometers, are installed on expensive devices where the cost of continuous condition monitoring is justified [12, 16].

However, vibration sensors are limited by their ability to detect electrical faults, for example in a motor stator. When continuously and remotely monitoring and diagnosing faults, it is necessary to check the operating status of the vibration sensors, which complicates the entire diagnostic procedure and increases the cost of its operation.

Electrical methods are without these disadvantages. Current diagnostics can be implemented on most machines using current sensors that are installed in the motor control unit. The use of current signals is suitable for diagnosing a large number of disks in remote mode.

The following phases of the development of a system for the diagnosis of mechatronic objects are proposed.

- Study of MS and it's functioning. Determination of possible errors. Collection of data on failures of similar objects and analysis of their reliability.
- Construction of mathematical model MS and its analysis.
- Selection of diagnosis method.
- Selection of diagnostic parameters.
- Development of information processing and decision-making algorithms based on neural networks, fuzzy sets. and genetic algorithms.
- Selection and development of hardware diagnostics.
- Development of diagnostic software.
- Testing and debugging the diagnostic system.
- Optimization of the diagnostic process.

The neural network with the reverse error propagation algorithm is used for diagnostics of MS. Errors in machining centers and CNC machines, located in the following subsystems, are considered as an example of errors in MS.

The mechanical subsystem's defects are in these nods:

- racks, shackle, trolleys, girder, columns, sliders, tables;
- universal spindle units, crankshafts, and driveshafts;
- gearboxes, starter clutches, brake clutches;
- cooling systems, lubricants;
- other parts and components.

Defects of electrical and electromechanical subsystems are located in the following nodes:

- propulsion engines, generators;
- electrical cabinets, electrical equipment;
- other elements of the subsystems.

Hydraulic (tire) defects are in the following nodes:

- hydraulic cylinders (pneumatic cylinders),
- hydraulic engines;
- hydraulic pumps;
- control devices.

The possible errors of CNC technological equipment and tools. To measure diagnostic parameters, force sensors are used to diagnose instrument, temperature sensors, current and voltage sensors for electric motors, sensors, and vibration sensors.

Figure 3.1 shows a block diagram of a hybrid intelligent diagnostic system that represents a software and hardware complex. The knowledge database (rules) contains a set of rules "If … then … ". The decision subsystem uses the knowledge database to process the information, the database contains.

The decision subsystem provides the interface with the operator and operation in real-time. The hybrid intelligent diagnostic system software is based on algorithms for processing the information and deciding the state/status of MS elements and

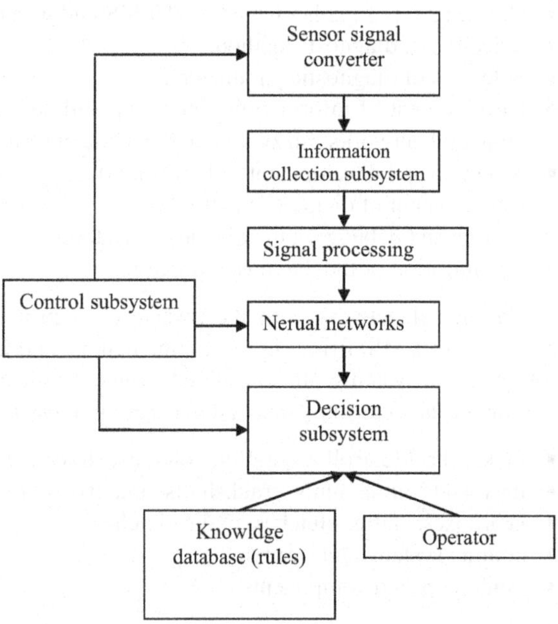

Fig. 3.1 Block diagram of a hybrid intelligent diagnostic system

nodes. Algorithms are based on intelligent data analysis algorithms [18, 20]. Such algorithms, called Data mining, enable the determination of the technical state and predict its change by the following tasks:

- The simulation of complex non-linear variations between input and output data,
- The identification of trends in forecasting data changes,
- Operating with loud signals and incomplete data,
- The actualization of the model, when new data becomes available,
- The identification of abnormal data.

To solve the problems of intelligent diagnostics, it is necessary to integrate neural networks with fuzzy logic. For example, in the diagnosis of the machine spindle unit, axial and radial loads and speeds have a large impact on vibration and noise. Under the influence of load, the bearing clearance/voltage, the stiffness, and the temperature in the bearing. The bearings clearance/voltage and the thermal deformation of the spindle is affected by the temperature change. Therefore, four parameters: axial and radial load, temperature, and rotation speed affecting the rules of using neural networks—x_1, x_2, x_3, x_4. Each rule uses its neural network. Following statements can be used:

- For example, you can use the following terms for loads:
  - No-load;
  - Low load;
  - Medium load;
  - Heavy load;
  - Extra heavy load.

Generally, we can write: "If $X = \{x_1, x_2, x_3, x_4\}$ is As, then ys is the output belonging to the neural network "S", where As is the fuzzy set of the conditional part of each rule. Each neural network S has n inputs for diagnostic parameters and own scales. Figure 3.2 shows a model of a neural network of fuzzy inference.

Based on the model of the neural network of fuzzy inference, the following algorithm for deciding on the technical state of MS is proposed.

Step 1. The creation of training and test samples. The training sample is created on a database acquired in different modes of MS operation.

Step 2. Cluster the training file. The training file is divided into r classes.

The N-dimensional entrance space is divided into r subspace. A decisive rule is set for each subspace.

Step 3. Training of the neural network defining the decision rule. For each input vector $X_i \in R_s$ there is a subspace for decision rule M_i. R_s is a subspace for decision rule. After training and testing, the neural network can determine the level to which each input vector belongs and to which class of subspace Rs.

In this algorithm, it is possible to modify membership functions as a result of obtaining new data from experts to obtain more reliable results.

Step 4. Neural network S learning. The training set with input vector $X_i \in R_s$ and output value y_s is fed to neural network S, a model of neural network.

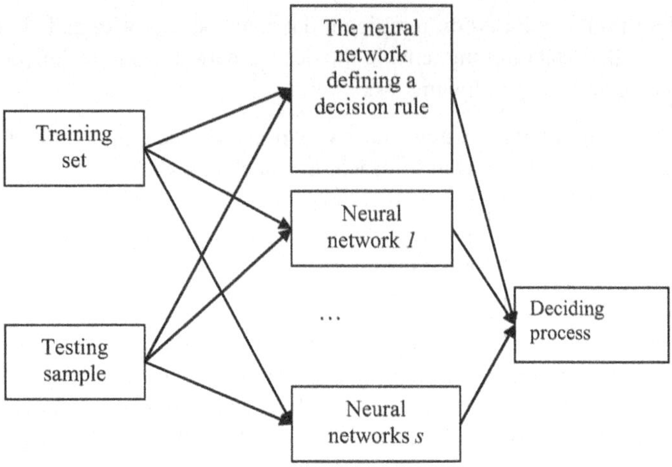

Fig. 3.2 Model of a neural network of fuzzy inference

Step 5. Decision making.

For a given input vector using neural network S, the output value is calculated.

Thus, using the presented algorithm, which uses a fuzzy inference neural network model, the technical state of MS elements and nodes is determined. On the input layer of the neural network, sensor signals inform about the modes of operation of the mechatronic system. In this layer, the number of neurons is determined by the number of diagnostic parameters. In the hidden layer, the number of neurons is usually equal to the number of neurons in the previous layer. The choice of the type and structure of the neural network is determined by the specifics of the problem being solved, but at present, there are no rules for network selection, types of neurons, number of layers in the network, and number of neurons in layers. Rosenblatt's perceptron network is suitable because it has an error propagation algorithm and allows us to minimize multilayer perceptron error. The neural network is taught with the help of a teacher. The presence of associative memory is not necessary because it does not matter whether the network remembers connections between neurons or not. The desired result can be achieved on a black-box model. To select the type of neurons, it is necessary to estimate the time to calculate the threshold activation function and the ability to differentiate this function. At each iteration of the backbone network, scales of the neural network are modified to improve the solution of one example.

For a given input vector using neural network S, the output value is calculated.

Thus, using the presented algorithm, which uses a fuzzy inference neural network model, the technical state of MS elements and nodes is determined. On the input layer of the neural network, there are sensor signals and information on the modes of operation of the mechatronic system. In this layer, the number of neurons is determined by the number of diagnostic parameters. In a hidden layer, the number of neurons is usually equal to the number of neurons in the previous layer. The choice of the type and structure of the neural network is determined by the specifics of the problem being solved. There are currently no network selection rules, types of neurons, number of layers in the network, and number of neurons in layers.

Rosenblatt's perceptron network is suitable because it has a back-propagation algorithm to minimize multilayer perceptron error.

The teacher is the performer of the teaching process of the neural networks.

Neural network teaching is performed with the help of a teacher. The presence of associative memory is not necessary, because whether the network remembers connections between neurons or not, is not important. The desired result can be achieved in the black-box model. To select the type of neurons, it is necessary to estimate the time to calculate the threshold activation function and the ability to distinguish this function. At each iteration of the spinal network, the weights of the neural network are modified to improve the solution of one example. Thus, in the learning process, the tasks of optimizing individual criteria are solved cyclically. In order to use the error backpropagation method, the neuronal transfer function must be differentiable, so exponential sigmoidity is chosen as the activation function.

Thus, in the learning process, the tasks of optimizing individual criteria are solved cyclically. In order to use the reverse error propagation method, the transmission function of neurons must be differentiable. Therefore, exponential sigmoidity is selected as the activation function. It is necessary to select the number of neurons and layers. If there are too few neurons or layers in the network, the network will not be able to learn, and the error will remain large during network operation. If there are too many neurons or layers, the network speed will be low, a lot of memory will be needed, the network will be retrained. This means that the output vector will transmit insignificant and irrelevant details on the output, e.g. will not be able to learn. Scaling is used to prepare the input and output data to bring the data to an acceptable extent. There is no general rule on how many hidden layers should be. Usually, there are 1–3 hidden layers. The more nonlinear problems, the more hidden layers should be. The speed of the neural network when processing diagnostic information depends on the number of neurons in the neural network layers. In a neural network, generally, all elements of the preceding layer are associated with all elements of the next layer. The number of neurons in the first layer depends on how many informative parameters are used for diagnostics. The number of neurons in the last layer depends on the result, to be obtained. The numbers of output neurons may be equal to the number of node states.

Using the neural network, the following stages of problem-solving of mechatronic systems are emphasized:

- Data selection for trainings;
- Data preparation and normalization;
- Choice of network topology;
- Experimental selection of network characteristics;
- Experimental selection of learning parameters;
- Proper network training;
- Checking the adequacy of training;
- Parameter settings;

- Final training;
- Publishing the network for future use.

3.2 Example of a Neural Network for Bearing Diagnostics

As an example, the development of a neural network for bearing assemblies diagnostics that are common nodes in MS. The Rosenblatt perceptron network with the Widrow-Hoff teaching algorithm was chosen for the diagnostics of MS bearing assemblies. This network allows minimizing multilayer perceptron error.

Four layers were chosen for the bearing assemblies diagnosis experiment: one input, two hidden layers for the weight attribution, and calculate the output parameters taking into account weights, one output. Then, after learning the network, three layers will be used:

1. input,
2. hiding and,
3. output.

Informative parameters of the bearing unit are input—it is the frequency and amplitude of vibrations determined by spectrum. The error frequency of bearing assemblies' various elements was given by known terms. The input parameter can be the temperature of bearing circles/rings, but the temperature can be increased by relatively serious errors. In the case of initial errors, the temperature of bearing assemblies does not increase, therefore it is not considered as a diagnostic feature in this example [7].

The Rosenblatt perceptron Network is a normal perceptron having a training sample consisting of input vectors, each having its own target vector. Components of input vector are represented as the continuous scope of values; the target vector components are binary values (0 or 1). After training, at the input, the network receives a set of continuous inputs and generates the required output as a vector with binary components. As is widely known, a defect in various states does not need to be a unique value but is within a range that requires a continuous field of output target values.

Firstly, layers are calculated in the last layer—layer A (based on output and reference signals), then in the penultimate layer—layer S, and after that in the input layer—layer X.

The maximum frequency of the vibration acceleration signal for single row radial ball bearing 6-180605 with double seals is 850.9 Hz. Table 3.1 shows the frequency of manifestations of various bearing defects.

For experiments, the MMA6233Q sensor with a frequency range up to 900 Hz was chosen, which has a built-in amplifier, low-pass filter with high sensitivity and a wide range of accelerations. During the experiments, the PCS500 digital oscilloscope was used to acquire an online frequency spectrum. The sensitivity of the MMA6233Q sensor is 120 mV/g. On the PCS500, the 1-volt digital oscilloscope

Table 3.1 Frequency of manifestation of various bearing defects

Designation of frequencies, where defects of a bearing are displayed	Name of frequencies	Values of frequencies of display of various defects of a bearing, Hz
f_{cage}	Rotational frequency of bearing cage	98.18
f_{re-or}	Frequency of moving of rolling elements along outer race	589.09
f_{re-ir}	Frequency of moving of rolling elements along inner race	850.9
f_{re-rot}	Rotational frequency of rolling elements	638.18

display corresponds to a vibration acceleration of 8.33 g. The DREMEL 300 electric drive was used as the motor for turning the inner ring of the bearing.

Figure 3.3 shows a photograph of an experimental test bench consisting of an electric drive (1), a bearing (2), a vibration acceleration sensor (3), a digital oscilloscope (4), a personal computer (5).

Figure 3.4 shows the error-free carrier signal spectrum 6-180605—of rotational speed error at a rotational speed of 11250 min^{-1}. The figure also shows that there are clear peaks at 100 and 200 Hz. These frequencies are harmonics of the industrial frequency of 50 Hz and are caused by interference from the 50 Hz network. There are also peaks at a frequency of 75 Hz and a second harmonic frequency of 150 Hz. Other

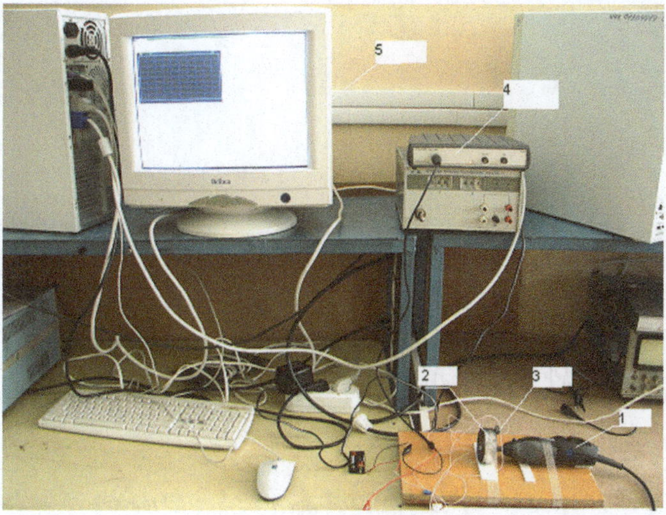

Fig. 3.3 Experimental stand

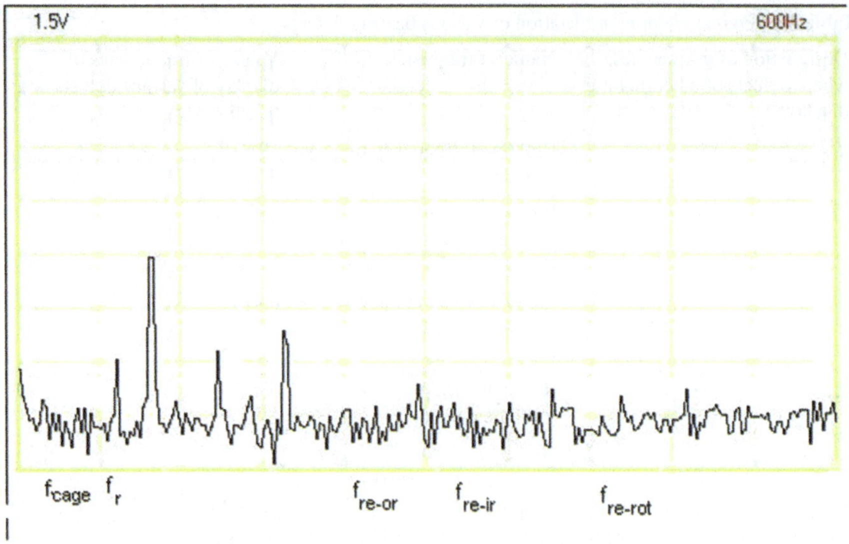

Fig. 3.4 The error-free carrier signal spectrum 6-180605 at 11,250 rpm

apparent vibration peaks are not observed, the average vibration level is negligible and is 0.5 V, indicating good bearing condition.

Figure 3.5 demonstrates the signal spectrum of the same bearing with artificial damage in form of transverse grooves on the outer ring. Defects of this type are reflected in the spectrum as peaks in the high-frequency field. By comparing damaged

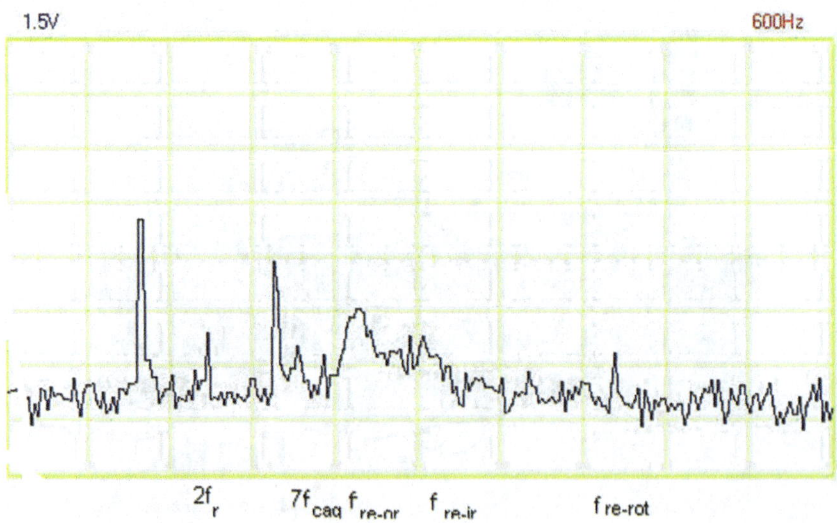

Fig. 3.5 The 6-180605 carrier signal spectrum 6-180605 at 11,250 rpm with damage

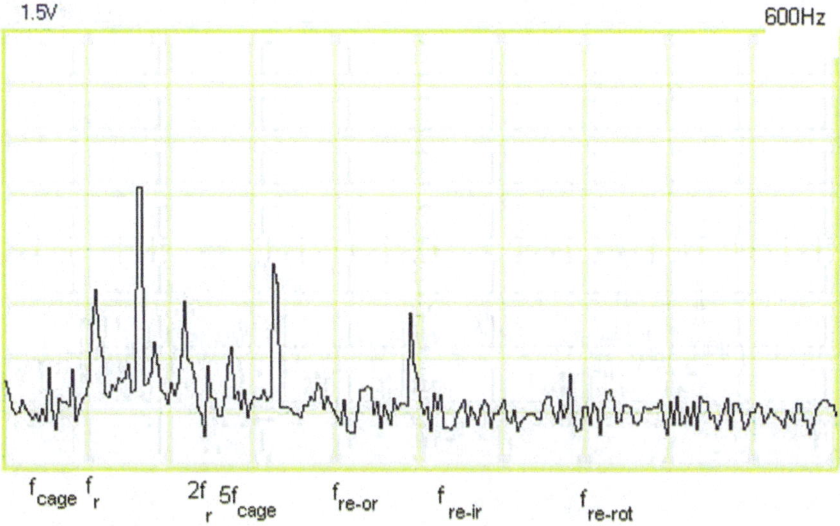

Fig. 3.6 The carrier signal spectrum 6-180605 at 11,250 rpm under radial load

bearing with damage-free bearing was discovered that the spectrum of the damaged bearing increases overall vibration level by 0.7 … 0.8 V and there are wide peaks at 250 Hz, which corresponds to the rolling frequency of anti-friction elements on the outer ring and 340 Hz, which corresponds to the rolling frequency of anti-friction elements on the inner ring.

Figure 3.6 illustrates the spectrum of bearings for defect 6-180605 with an increased radial load. In this graph, the amplitudes of the spectral components increased at frequencies that are multiples of 50 Hz, at a separator rotation frequency of 0.78 … 0.8 V at the rotor frequency and at its second harmonic frequency.

Figure 3.7 shows the signal spectrum of bearing 6206 at 7800 rpm without error. This spectrum also shows peaks at a frequency that is a multiple of 50, 100, and 150 Hz. The average vibration amplitude corresponds to a level of 3 V.

The spectrum of the same bearing with a defect is demonstrated in Fig. 3.8. The defect is damage to the outer ring in the form of transverse furrows. When comparing the defect-bearing spectrum with the defect-free bearing spectrum, an increase in vibration amplitude of 0.7 … 0.8 V is observed at a frequency of 270 Hz, which belongs to the anti-friction element rolling frequency along the inner ring. The harmonic rotational frequencies of the bearing cage and outer ring will be recorded.

The signal spectrum of bearing 6206 without damage at 14,400 rpm is illustrated in Fig. 3.9. This spectrum also shows peaks at a frequency that is a multiple of 50, 100, and 150 Hz. The average vibration amplitude corresponds to a level of 3 V.

Figure 3.10 illustrates the spectrum of the same bearing as in Fig. 3.9 but with damage. When comparing the defect-bearing spectrum with the defect-free bearing spectrum, an increase in the overall vibration level is observed at frequencies from 360 Hz to 0.7 … 0.8 V, with the maximum peak decreasing to the 510 Hz, the rolling

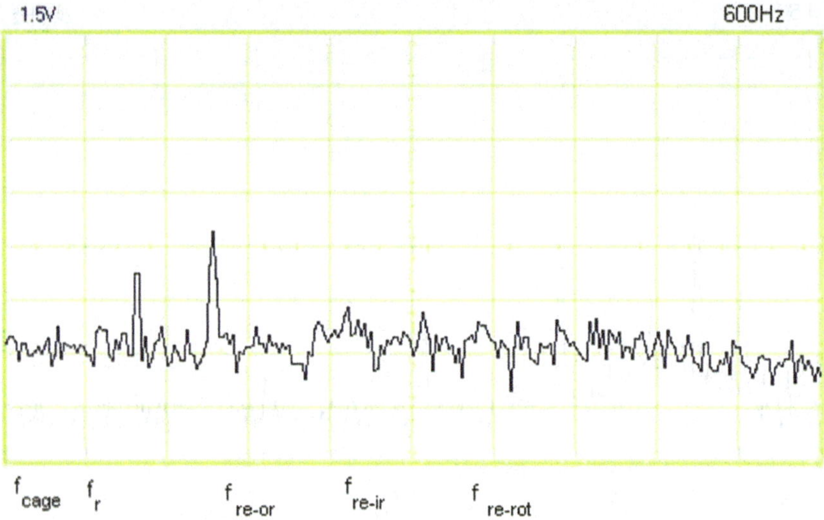

Fig. 3.7 The carrier signal spectrum 6-180605 at a rotational speed of 180 rpm without error

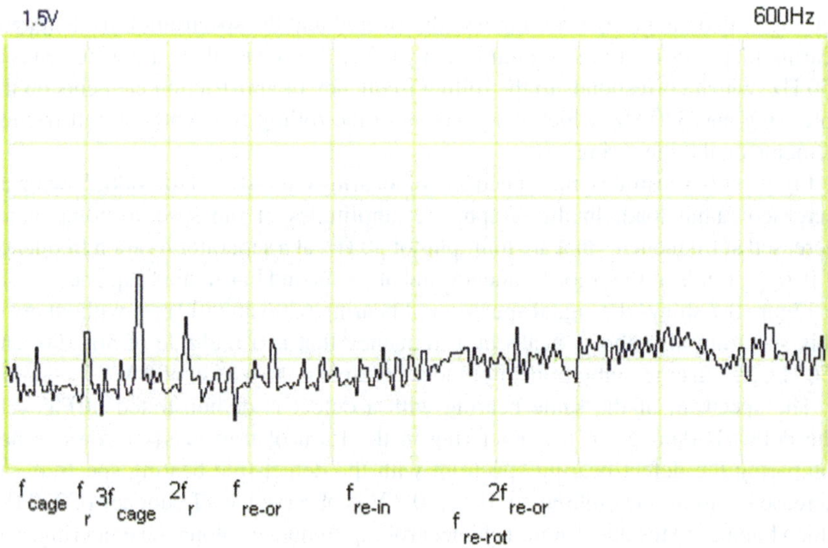

Fig. 3.8 The 6-180605 carrier signal spectrum at 7800 rpm with damage

frequency of the outer ring anti-friction elements. The amplitude increased by 2 … 2.5 V.

When errors are present, the amplitude increases by an average of 0.75 V throughout the frequency spectrum, the vibration level increases by 1.5 … 2 V at the

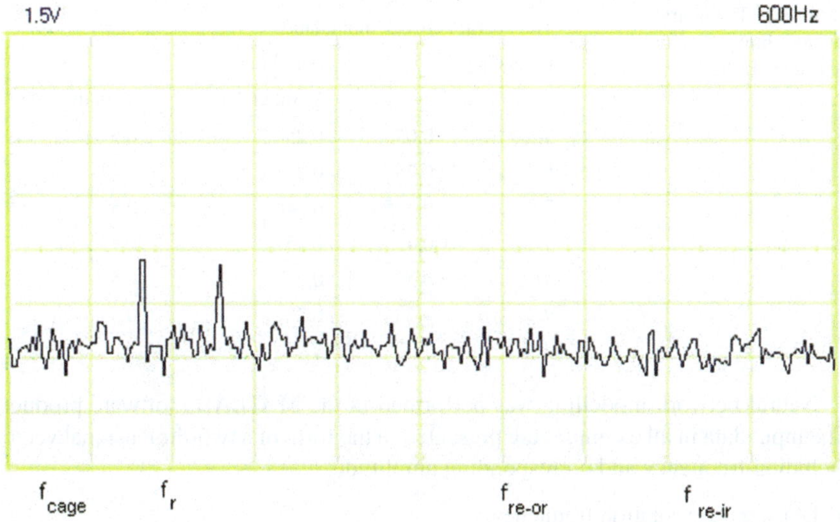

Fig. 3.9 The carrier signal spectrum 6-180605 at 14,400 rpm without damage

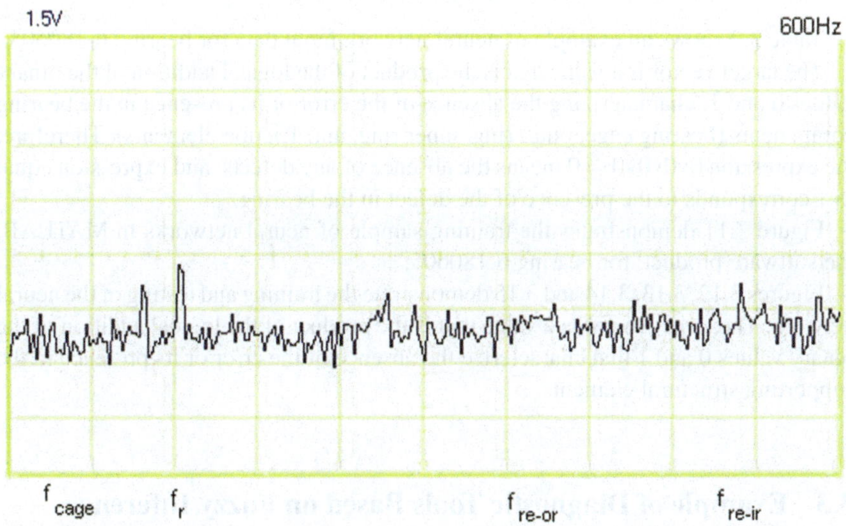

Fig. 3.10 The carrier signal spectrum 6-180605 at 14,400 rpm with error

rotational frequencies of the outer and inner ring anti-friction elements. While under the radial load the amplitude increases at rotor speed and at the industrial frequency of 50 Hz the peaks appear at the rotational speed of the separator.

Table 3.2 The neural network input data for the bearing 6-180605

The neural network input data			
Frequency		Amplitude	
		Without defect	With defect
f_{cage}	0.05	0.25	0.5
$5f_{cage}$	0.25	0.21	0.47
$7f_{cage}$	0.375	0.246	0.53
f_{re-or}	0.42	0.25	0.75
f_{re-ir}	0.55	0.25	0.5
f_{re-rot}	0.73	0.25	0.33

Neural network modelling was performed in the MATLAB (software product). The input data in all examples are presented in the form of a two-dimensional vector, including frequency and corresponding amplitude:

- bearing cage rotation frequency,
- the bearing balls noise frequency on the outer ring,
- the bearing balls noise frequency on the inner ring,
- the rotational frequency of the anti-friction elements.

Table 3.2 shows an example of neural network input data for bearing 6-180605.

The target vector is a value that is the product of the logical addition of the binary values 0 and 1, characterizing the absence of the error or its presence in the bearing components (bearing cage, outer ring, inner ring, anti-friction elements). Therefore, the expression $0\text{v}0\text{v}0\text{v}0 = 0$ means the absence of any defects, and expression equal to 1 corresponds to the presence of the defect in the bearing.

Figure 3.11 demonstrates the training sample of neural networks in MATLAB, the software product, for bearing 6-180605.

Figures 3.12, 3.13, 3.14 and 3.15 demonstrate the training and testing of the neural network. The target vector is a value that is the product of the logical addition of the binary values 0 and 1 that characterize the absence of the error or its presence in the supporting structural elements.

3.3 Example of Diagnostic Tools Based on Fuzzy Inference Systems

The fuzzy model based on deductive derivation was developed for machine diagnostics. It is caused by the fact that the uncertainty, the lack of information must be faced when dealing with this problem. To solve this problem, it is necessary to determine the input and output parameters of the diagnostic system, the matrix of binary relations. The diagnostic system should determine the input parameter values (machine condition) with the output parameter values (diagnostic parameters). The

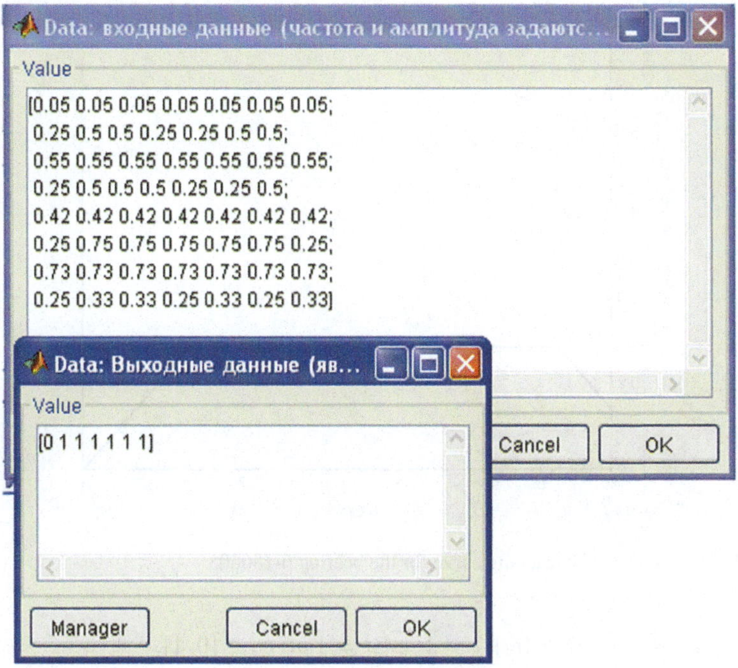

Fig. 3.11 The training of neural network sample of the bearing 6-180605

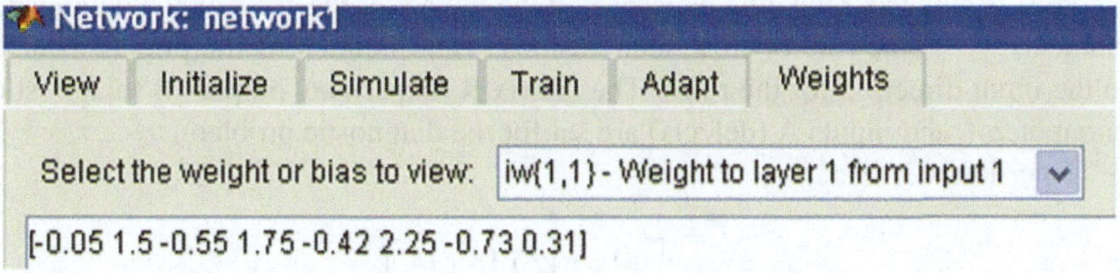

Fig. 3.12 The scales of neural network for the bearing 6-180605

input parameter field is defined by machine failure X vector (m = 5) and total output parameter—diagnostic vector Y (n = 4).

$$X = \{x_1, x_2, x_3, x_4, x_5\} \tag{3.1}$$

$$Y = \{y_1, y_2, y_3, y_4\} \tag{3.2}$$

There are fuzzy causal connections between x_i and y_j. In the example, fuzzy relationships are represented as a 5-row and 4-column matrix R, i.e. fuzzy relationship matrix.

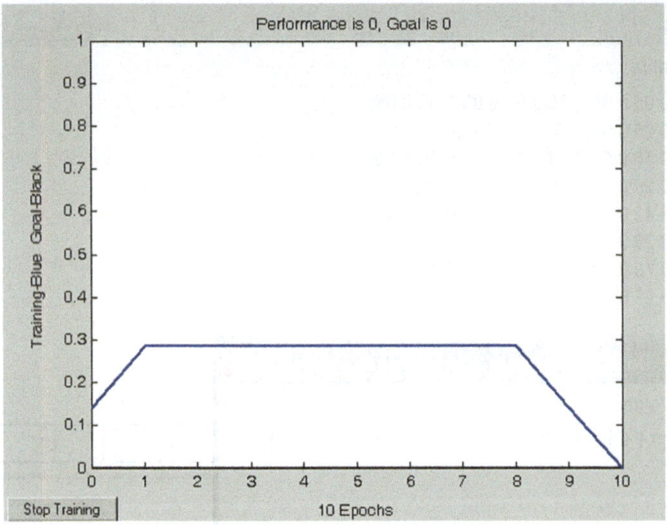

Fig. 3.13 Neural network teaching curve for the bearing 6-180605

$$\mathrm{R} = [\mathrm{r_{ij}}];\ i = \overline{1, m};\ j = \overline{1, n};\ r_{ij} \in [0, 1]. \tag{3.3}$$

The specific inputs and outputs of the system are considered to be fuzzy sets A and B in space of X and Y. The connections of these sets are denoted as B = A ∘ R, where R is a matrix reflecting expertise on the impact of the error on the diagnostic element; "∘" is the rule of fuzzy conclusions. The direction of the pins is inverse to the input direction for the rules. The matrix R (expertise), outputs B (diagnostic parameters), and inputs A (defects) are set for the diagnostic problem.

Expertise is:

$$R = \begin{bmatrix} r11\ r12\ r13\ r14 \\ r21\ r22\ r23\ r24 \\ r31\ r32\ r33\ r34 \\ r41\ r42\ r43\ r44 \\ r51\ r52\ r53\ r54 \end{bmatrix} \tag{3.4}$$

In matrix (3.4), the first column corresponds to the diagnostic parameter y_1; the second column is diagnostic parameter y_2 the third column is for diagnostic parameter y_3, the fourth column is for diagnostic parameter y_4. In this case, for example, the diagnostic column y_1 is caused by the first column of X defects. Values of the degree of conformity are similar to classical probabilities, but it does not require that the sum must be equal to one, as in classical probability theory. Conformity level values are similar to classical probabilities, but this is not a precondition for the solution

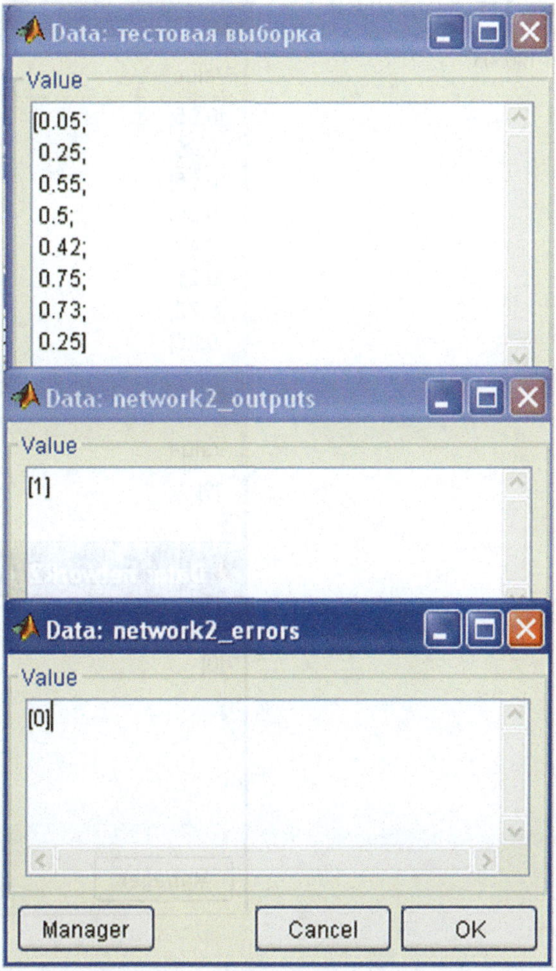

Fig. 3.14 Neural network results for the bearing 6-180605

because $\sum_{i=1}^{m} \mu_i = 1$, i.e. the amount does not need to be the same as in the classical probability theory.

Assuming that the result of measuring diagnostic parameters of machine condition is estimated by specialists as:

$$B = \frac{b_1}{y_1} + \frac{b_2}{y_2} + \frac{b_3}{y_3} + \frac{b_4}{y_4}, \tag{3.5}$$

i.e. diagnostical characteristics, y_j, $j = \overline{1, 4}$, that is performed with conformity level $b_j = \mu_j$, $j = \overline{1, 4}$.

The technical condition of the machine is determined by the formula:

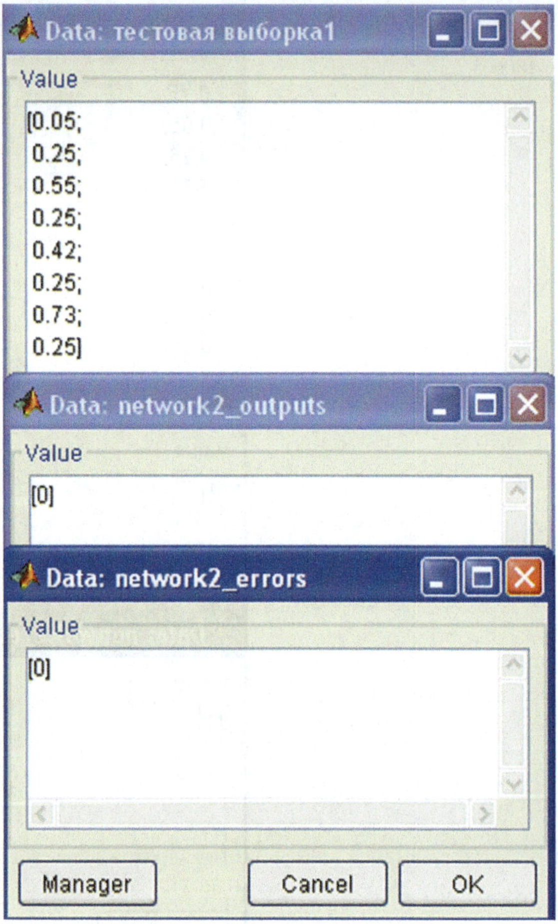

Fig. 3.15 Neural network results for the bearing 6-180605

$$A = \frac{a_1}{x_1} + \frac{a_2}{x_2} + \frac{a_3}{x_3} + \frac{a_4}{x_4} + \frac{a_5}{x_5}. \tag{3.6}$$

The formulas for vectors B and A are presented as chains:

$$B = [b_1; b_2; b_3; b_4], \tag{3.7}$$

$$A = [a_1; a_2; a_3; a_4; a_5], \tag{3.8}$$

where $b_j, a_i \in [0, 1]$; $j = \overline{1, 4}$; $i = \overline{1, 5}$.

The formula B = A ∘ R is represented as:

$$
[b_1; b_2; b_3; b_4] = [a_1; a_2; a_3; a_4; a_5]^{\circ} \begin{bmatrix} r11 & r12\,r13\,r14 & & \\ r21 & r23 & r23 & r24 \\ r31 & r32 & r33 & r34 \\ r41 & r42 & r43 & r44 \\ r51 & r52 & r53 & r54 \end{bmatrix}, \tag{3.9}
$$

The "product" of vector A and the matrix R is calculated, but a minus subtraction operation ("∩"—min) is performed instead of a multiplication operation and a maximum sensing operation ("U"—max) is performed instead of an add operation. In general, there is a single maximum and several "smaller" solutions for the composition of the absolute minimum. The solution is, therefore, a vector of values, each of which belongs to a particular segment (within the interval 0–1).

Quite often, due to the manifestation of the subjective human factor (minor professional errors) in the assessment of the condition, it is not possible to obtain an accurate solution for the system B = A ∘ R. In this case, the closest (minimum sum of absolute deviations from zero for each of the equations of the above system) is found to be the approximate solution (or solution) of the system B = A ∘ R, and the approximate solution obtained is accepted or rejected. If the overall deviation is large enough (from the expert's point of view), the expert should be asked again to evaluate the technical condition, as it is clear that he introduced conflicting estimates of the technical condition during the first assessment.

An example of a program for determining the technical condition of a CNC machine, given in Appendix A, is written in the C++ programming language in the MFC development environment. A significant advantage of MFC is the simplified interaction with the Windows application programming API. After starting the program, a dialog box will appear, through which the program communicates with the user. The user enters an expert matrix, selects one of the analysed parameters. Basic values can range from 0 to 1.

The program algorithm can be represented as a sequence of steps.

- ***Step 1***. Setting or correction of initial data [matrix R of binary relations between the values of the output parameters B (diagnostic parameters) and the values of the input parameters A (technical condition)] (Fig. 3.16).
- ***Step 2***. Entering measured diagnostic parameters—vector B.
- ***Step 3***. Finding vector A (presence of errors) to solve the system of equations B = A ∘ R, where the disjunction is replaced by a maximum and the conjunction is replaced by a minimum.

$$
b_1 = (r_{11} \wedge a_1) \vee (r_{12} \wedge a_2) \vee (r_{1m} \wedge a_m)
$$

$$
\ldots
$$

$$
b_n = (r_{n1} \wedge a_1) \vee (r_{n2} \wedge a_2) \vee (r_{nm} \wedge a_m)
$$

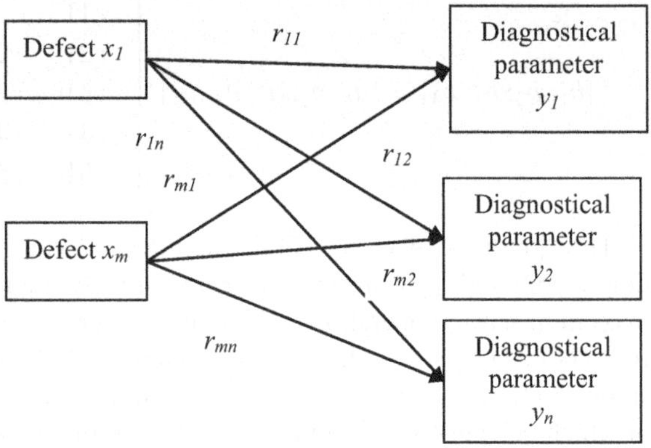

Fig. 3.16 Diagram of the impact of defects on diagnostic parameters

- ***Step 4.*** Obtaining solution A in the form of an interval in which boundaries define the minimum and the maximum solutions.

3.4 Example of Diagnostics of Mechatronic Dynamic Modules

To develop a system for diagnosing mechatronic dynamic modules (MDMs), MDM errors are detected and patterns between errors, modes of operation, and diagnostic parameters are analysed. Based on these laws, the fuzzy logic rules base is established to determine the technical state of the MDM. The model for determining the technical condition of MMD is developed in the program MVTU (modelling in technical equipment, buildings, and systems) developed at Bauman Moscow State Technical University.

For all measured diagnostic parameters, the allocation is in the range of −1 to 1. The fuzzification of the diagnostic parameters and movement speed is performed using the Gaussian curve member function. For each diagnostic parameter and speed, three terms are defined that are equally divided in range −1 and 1. The fuzzy rule conclusions are constructed as conditional operators with weights/scales for each rule. Numerical integration determines the output quantity—technical state level (−1—good technical state (without errors), 1—significant errors in MDM). To create rules for fuzzy inference, a table is created in which the logical operations between the input parameters are AND operations ("AND"). Table 3.3 gives an example of three diagnostic parameters and speed.

Terms for the current technical state of MDM:

L—error-free; M—with small errors; H—with significant errors.

Examples of model and simulation results at different speeds are shown in Figs. 3.17, 3.18 and 3.19.

The block diagram of an intelligent MDM with a self-diagnostic subsystem is shown in Fig. 3.20.

Table 3.3 Defect dependence on MDM parameters and speed

No. conditions	Temperature	Vibration	Current	Speed	Defect appearance
1	L	L	L	H	L
2	M	M	M	M	M
3	H	H	H	L	H

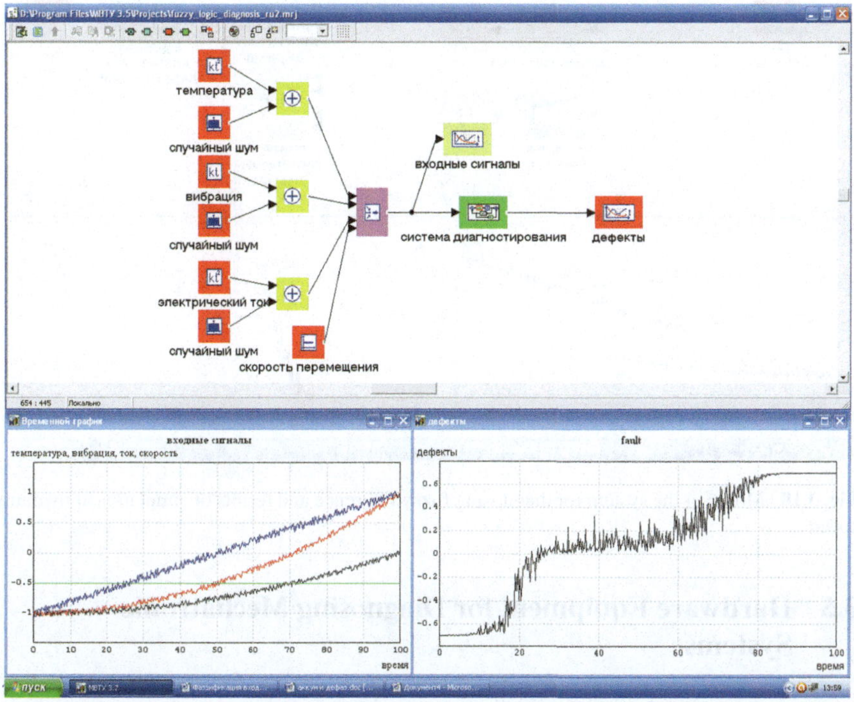

Fig. 3.17 Model of fuzzy inference system for diagnostics and simulation results

If the MDM state is defect-free, then the monitoring is performed using the fuzzy PID controller. For a fuzzy PID controller, fuzzification is performed by the proportional, integral, and differential components of the mismatch error using the fuzzy sets. If the MDM has been defected, then during performing the operation will be considered the degree of development of these errors and the forecast of the possibility of meeting the control objective. Information about defects is transmitted to the operator and to a higher level of control.

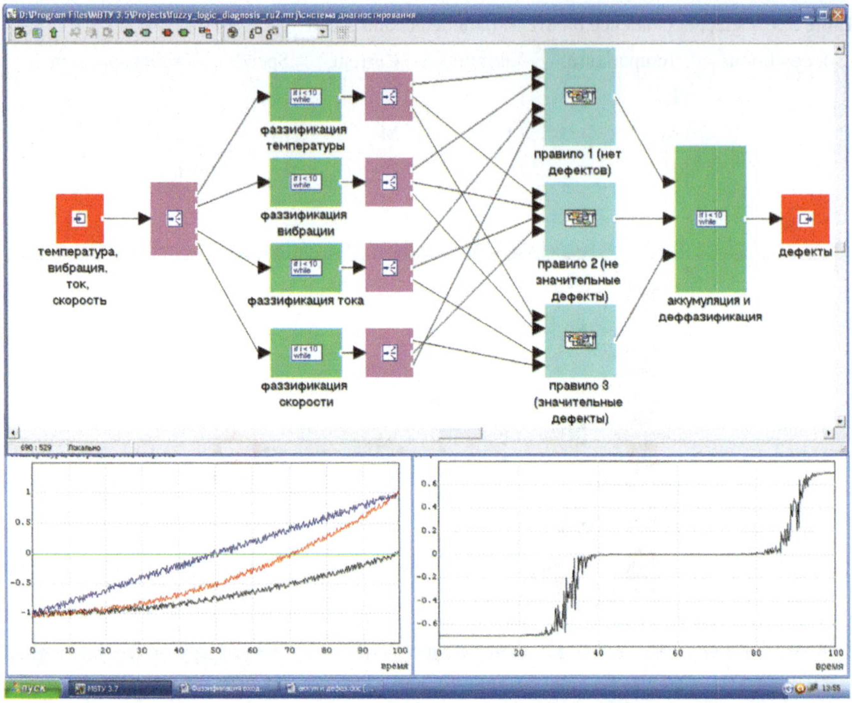

Fig. 3.18 Model of the system for diagnosing fuzzy inference and results of simulation at medium speed

3.5 Hardware Equipment for Diagnosing Mechatronic Systems

Mechatronic systems must be competitive in terms of quality and cost. This imposes certain limitations on the hardware and software of the diagnostic system [8, 19].

Depending on the organization's diagnostic method, MS hardware diagnostic models are grouped into three groups: parallel, sequential, combined.

In a parallel array, the collection, processing of sensor information, and the technical state decision of the MS are made in parallel with the computing devices found in each mechatronic module. The computing devices can be microcontrollers or digital signal processors that transmit the solution results to the local MS network.

A CNC or PC can be connected to the LAN network. Sensors are located at the point where diagnostic signals are generated. A schematic diagram of the parallel diagnostic device is shown in Fig. 3.21. With the MS Diagnostic Sequence Organization, the collection, processing of sensor information, and technical status decisions are made using a single computing device, which may be a microcontroller, digital signal processor, or industrial computer [17]. Sensors are located at the point where diagnostic signals are generated. A scheme of the serial diagnostics of the device

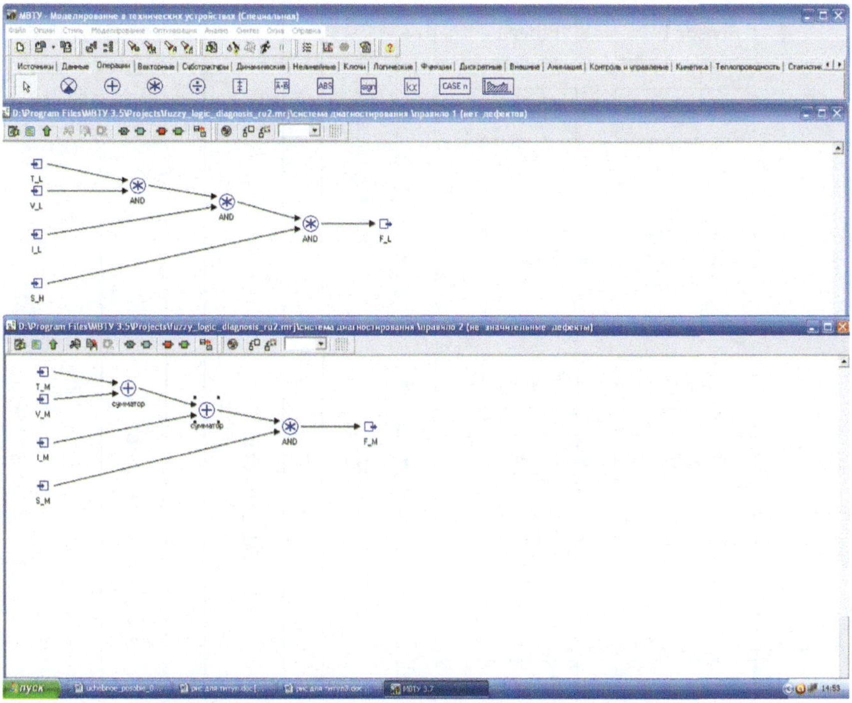

Fig. 3.19 Fuzzy derivation of model rules in the absence of defects and minor errors

Fig. 3.20 The block diagram of an intelligent MDM with a self-diagnostic subsystem

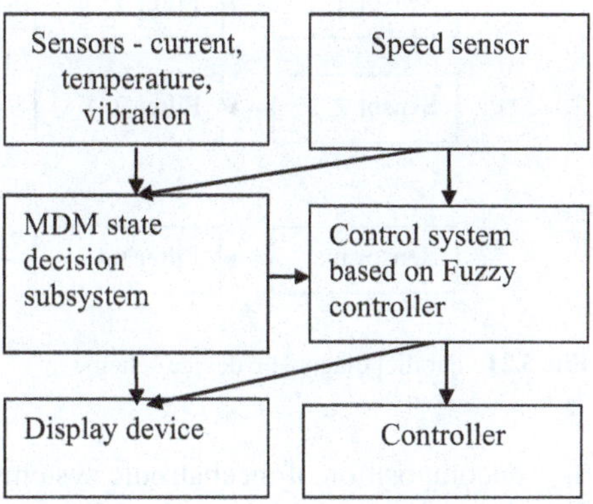

is shown in Fig. 3.22. With the combined grouping of MS diagnostics, the most important and critical modules are diagnosed in parallel and the remaining modules are diagnosed sequentially. The combined device diagnostics scheme is shown in Fig. 3.23.

The following algorithm is designed for the construction of devices for diagnostics of mechatronic systems, which consists of a sequence of the following steps:

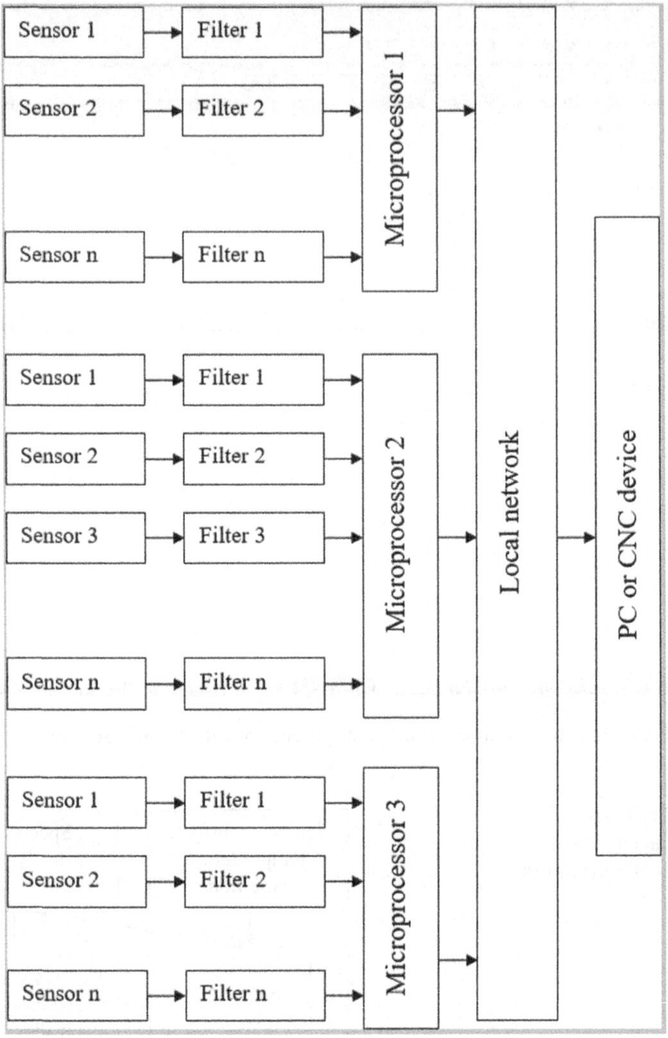

Fig. 3.21 Parallel diagnostic device scheme

1. decomposition of mechatronic systems into modules, nodes, elements;
2. determination of diagnostic parameters in modules, nodes, elements;
3. selection of sensors to measure diagnostic parameters;
4. selection of diagnostic intervals.

The algorithm for the construction of MS diagnostic systems is considered on the example of a CNC machine. A CNC machine, such as MS, consists of mechanical, electrical, electromechanical, electronic subsystems, CNC equipment. Presence of hydraulic and pneumatic subsystems.

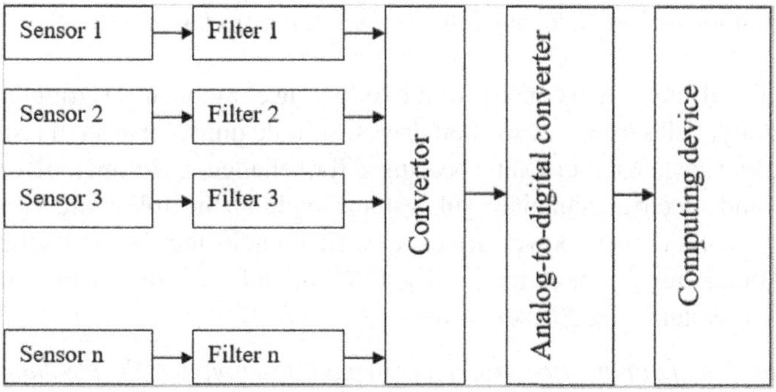

Fig. 3.22 Serial diagnostics of the device

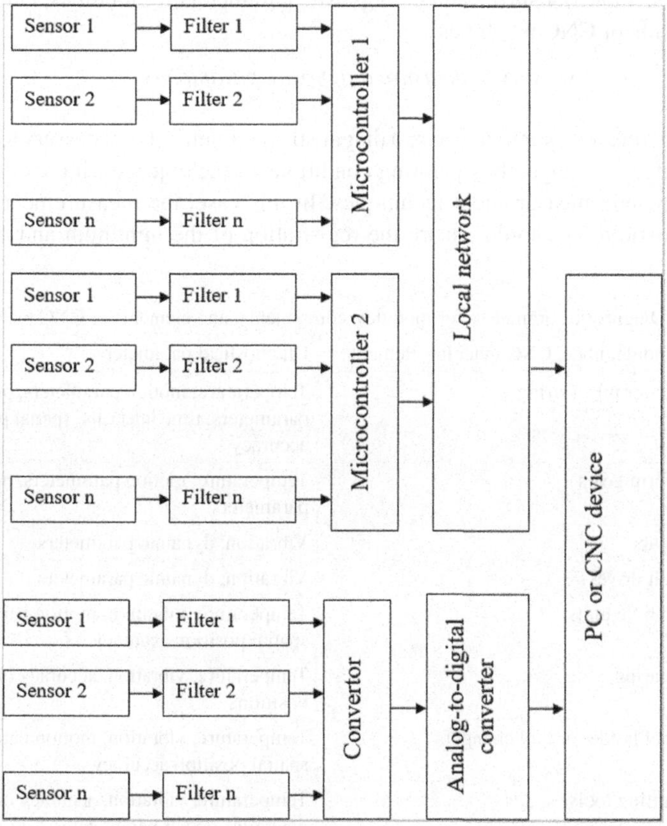

Fig. 3.23 Diagram of the combined diagnostic equipment

(1) *Decomposition of CNC machine into modules, nodes, elements*

The mechanical subsystem consists of the following components: setting/bearer pile, caliper, trolley; ball screws; gears, belt drives; spindle units, drive shafts; speed, feed boxes; cooling systems, lubricants; bearings; Tool changers; cutting tool; other parts. Electrical and electromechanical subsystems include the following components: main propulsion engines, power drives, electrical enclosures with electrical equipment, and other subsystem elements. The CNC includes the following components: drive control systems, feedback sensors.

(2) *Definition of diagnostic parameters in modules, nodes, CNC machine elements*

In the paper [1] presents some diagnostic objects of CNC lathes and their diagnostic parameters. Table 3.4 shows some diagnostic parameters for modules, components, and elements of CNC machines.

(3) *Selection of sensors to measure diagnostic parameters*

When selecting a sensor to measure a diagnostic parameter, it is necessary to consider the measurement range, the operating conditions of the object during measurement, availability, and measurement techniques. In this case, the measurement range of the diagnostic tools should ensure the registration of the minimum and maximum

Table 3.4 Diagnostic parameters for modules, components, and elements of CNC machines

No.	Module, node, CNC machine element	Diagnostical parameters
1	Bearer pile a string	Temperature, motion parameters, power parameters, time intervals, spatial position accuracy
2	Worm gears	Temperature, motion parameters, power parameters
3	Gears	Vibration, dynamic parameters
4	Belt drives	Vibration, dynamic parameters
5	Spindle units	Temperature, vibration, motion parameters, spatial position accuracy
6	Bearings	Temperature, vibration, accuracy of spatial positions
7	Tool holder or tool changer	Temperature, vibration, motion parameters, spatial position accuracy
8	Cutting tools	Temperature, vibration, accuracy of spatial positions, power parameters
9	Electromotor	Current, voltage, power, temperature, vibration, movement parameters
10	Drive control systems	Current, voltage, power, temperature
11	Sensors	Motion parameters, time intervals
12	Poppet head	Temperature, spatial accuracy

Table 3.5 Diagnostic parameters for modules, components and elements of CNC machines and sensors for their measurement

No.	Diagnostic parameters	Sensors
1	Electrical current	Current sensors up to 100 A, operating frequency 0–25 kHz
2	Electrical voltage	Voltage sensors 10–500 V, operating frequency 0–25 kHz
3	Power	Power sensors 0.5–20 kW, operating frequency 0–25 kHz
4	Temperature	Temperature sensors 0–150 °C
5	Motion parameters	Accelerometers ± 2 g, encoders 10,000 pulses/rotations
6	Performance parameters	Tensile force sensors up to 10 kN
7	Time intervals	Timers in the controller
8	Vibration	Accelerometers ± 2 g, operating frequency 1–25 kHz
9	Spatial position accuracy	Encoders 10,000 pulses/rotation

(limit) values of the diagnostic parameters. The sensor measurement error should be 1–2%. If it is possible, all sensors, especially vibration and temperature sensors, shall be installed in the immediate vicinity of the diagnosed object. The presence of embedded sensors in machine elements and components is ideal, such as position sensors, angular velocity, temperature, and vibration, as well as a microcontroller to convert information to digital form, processing and transfer to other controllers, are built into mechatronic bearings. Table 3.5 provides diagnostic parameters and sensors to measure them.

(4) *Selection of diagnostic intervals*

The order of analysis of diagnostic parameters depends on the level of responsibility of functional elements of MS, on the time of their diagnosis, and on the probability of defects appearance therein. For example, in practice, the alignment of the functional element diagnostics sequence occurs in increasing order of the ratio of the time required to diagnose the functional element to the probability of failure of the functional element. The diagnostics interval of functional elements depends on the degree of responsibility of the mechatronic module, the node, the MS element, and the speed of degradation processes therein. A general criterion K_i is proposed which refers to the level of responsibility of the ith mechatronic module, the node, the MS element, and the rate of flow of degradation processes in them, calculated as:

$$K_i = K_{otv} + K_{\text{degr}}, \tag{3.12}$$

where K_{otv} is ith elements responsibility coefficient, ranging from 0 to 0.5 (0.5 is the maximum degree of responsibility); the K_{degr} is coefficient characterizing the flow rate of degradation processes of the i-th functional element ranges from 0 to 0.5 (0.5 is the maximum flow rate of degradation processes). A high K_i value means that more critical modules, nodes, elements with a high degree of degradation processes should be diagnosed more frequently. The above factors are determined by the expert estimation method. The approximate values of the coefficients are shown in the matrix K. The columns in the matrix are arranged in order of increasing rate of degradation of functional elements (the first column corresponds to the slow degradation rate of the object of diagnosis, the third column is high). The rows in the matrix are arranged to increase the responsibility of the functional elements.

$$K = \begin{matrix} 0,1\ldots0,3 & 0,4\ldots0,6 & 0,7\ldots1,0 \\ 0,2\ldots0,4 & 0,4\ldots0,6 & 0,6\ldots0,8 \\ 0,4\ldots0,6 & 0,6\ldots0,8 & 0,8\ldots1,0 \end{matrix}. \tag{3.13}$$

The diagnostic interval T is calculated according to the formula

$$T = T_c/K_i, \tag{3.14}$$

where T_c is the time of the diagnostic cycle determined by the hardware and software capabilities of the diagnostic equipment; K_i is a common criterion. Table 3.6 lists the diagnostic criteria and intervals for modules, components, and elements of CNC machines.

Diagnostic intervals are calculated in the last column of Table 3.6, where T_{min} is the minimum diagnostic interval. Calculates the sum of diagnostic intervals for modules, components, and elements of CNC machines. In our example, the sum of diagnostic intervals is 20.9 T_{min}.

The multiplicity of the diagnostic intervals is then calculated by dividing the sum to × T_{min}. The minimum multiplicity is 7.0 and the relative multiplicity is calculated—all values are divided by the minimum multiplicity of 7.0. For practical implementation, the relative multiplicity is rounded to integers. Table 3.7 shows

Table 3.6 Criteria and intervals for diagnostics of modules, nodes, and elements of CNC machines

No.	Module, node, element	K_{otv}	K_{degr}	K_i	T_i	$k \times T_{min}$
1	Setting and wiring	0.5	0.1	0.6	1.67	$1.5T_{min}$
2	Ball helix	0.4	0.2	0.6	1.67	$1.5T_{min}$
3	Cog-wheel	0.2	0.3	0.5	2.00	$1.8T_{min}$
4	Belt gears	0.2	0.3	0.5	2.00	$1.8T_{min}$
5	Spindle units	0.4	0.3	0.7	1.43	$1.3T_{min}$
6	Bearing	0.3	0.3	0.6	1.67	$1.5T_{min}$
7	Tool holder or tool changer	0.2	0.2	0.4	2.50	$2.3T_{min}$
8	Cutting tool	0.4	0.5	0.9	1.11	T_{min}
9	Electric motors	0.2	0.3	0.5	2.00	$1.8T_{min}$
10	Drive control systems	0.2	0.2	0.4	2.50	$2.3T_{min}$
11	Sensors	0.5	0.3	0.8	1.25	$1.1T_{min}$
12	Poppet head	0.1	0.2	0.3	3.33	$3.0T_{min}$

Table 3.7 Frequency of intervals for diagnostics of modules, nodes, and elements of CNC machines

No.	Module, node, detail	Failure rate	Relative failure rate	Average failure rate
1	Setting, slide	13.9	2.0	2
2	Ball helix	13.9	2.0	2
3	Cogwheel	11.6	1.7	2
4	Belt gears	11.6	1.7	2
5	Spindle units	16.1	2.3	2
6	Bearing	13.9	2.0	2
7	Tool holder or tool changer	9.1	1.3	1
8	Cutting tool	20.9	3.0	3
9	Electric motors	11.6	1.7	2
10	Drive control systems	9.1	1.3	1
11	Sensors	19.0	2.7	3
12	Poppet head	7.0	1.0	1

the frequency of intervals for diagnostics of modules, nodes, and elements of CNC machines.

For one diagnostic cycle, it is, therefore, necessary to diagnose for example setting and slides 2 times, cutting tool 3 times, poppet-head 1 time. As an example we propose a sequence of diagnostic modules, components and elements of CNC machines: 8, 11, 1, 2, 3, 4, 5, 8, 11, 6, 7, 9, 1, 2, 3, 8, 11, 4, 5, 6, 9, 10, 12.

3.6 Multicriterial Optimization of Diagnostic Systems

In order to solve the problem of optimizing the MS diagnostics process, it is necessary to select the criteria for optimizing the MS diagnostic devices. The selection of criteria for optimizing MS diagnostic devices is a relatively complex task, given the need to take into account a large number of factors with varying degrees of significance at the same time [11]. A generalized criterion of optimality of MS diagnostic equipment is defined as the functionality of economic, organizational, technological, and technical criteria.

$$Y = F(X_E, X_{OT}, X_T), \tag{3.15}$$

wherein Y is a general optimality criterion of MS; X_E is economic criteria; X_{OT} is organizational and technical criteria; X_T is technical criteria. Economic criteria include accident losses, maintenance and repair costs, scrap volume, especially in the manufacture of expensive products, use of working time (readiness factor), diagnostic costs, etc. It is clear that the above economic criteria should be optimized.

Optimization according to economic criteria is therefore also multi-criteria. The economic criterion of optimality—the economic efficiency of using a diagnostic system is defined as the functionality of private economic criteria.

$$X_E = F(X_1, X_2, \ldots, X_N), \tag{3.16}$$

where X_i is subjective economic criteria. The economic criteria are calculated as the difference in costs—for operating MS without using diagnostic equipment and for operating MS using diagnostic equipment. For example, the economic results of the use of diagnostic systems can be determined by the capital investment efficiency ratio, which expresses the annual savings from the use of diagnostics [3].

$$E = (C_1 - C_2)/(K_1 - K_2), \tag{3.17}$$

where C_1 and C_2 are the primary costs of annual production without diagnosis and diagnosis of MS condition;

K_1 and K_2 are capital expenditures for the production of the annual production of components without the use of the diagnostic system and with diagnostics of MS condition.

The organizational and technical criteria include: the procedure of analysis of diagnostic parameters, the accuracy of determining the technical condition, the time and the interval of diagnostics of functional elements of MS. Optimization according to organizational and technical criteria is also multicriteria.

The order of analysis of diagnostic parameters depends on the degree of responsibility of the functional elements of the MS, on the time of their diagnosis and on the probability of occurrence of defects in them [4, 6]. For example, in practice, there is an organization of a sequence of diagnostics of MS functional elements in increasing order of the ratio of the time required to diagnose a functional element to the probability of failure of a given functional element. The time interval for the diagnostics of functional elements depends on the degree of responsibility of the mechatronic module, the node, the MS element and the speed of the degradation processes in them. The time interval for the diagnostics of functional elements depends on the degree of responsibility of the mechatronic module, the node, the MS element and the speed of the degradation processes in them.

Optimization according to technical criteria is also multicriteria. Solving the problem of optimization for the process of diagnosing MS with regard to various criteria is considered to be a solution with several criteria: maximum economic efficiency of using the diagnostic system, maximum accuracy of technical condition, minimum diagnostic time, the minimum cost of diagnostic equipment, minimum weight and size [23].

There are two trends in multicriteria optimization. The first trend of multicriteria optimization, used mainly in insufficiently formalized areas of science, such as economics, is to organize a dialogue between the decision-maker and the computer. This trend is based on expert systems [5]. The second trend of multicriteria optimization, used in technical sciences, is the development of specialized computational methods covering a selected class of systems and specialized optimization criteria [4].

To solve the problem of optimization of MS diagnostics, it is appropriate to use the second trend of optimization of several criteria.

Currently, there are four approaches to the problem of multicriteria optimization—the problem of reduction to optimization to one criterion [4]:

1. *Creating a Pareto region*

and giving the decision to the decision-maker to choose only one Pareto-optimal solution. This approach is quite laborious because it provides a multidimensional space of criteria with different dimensions and it is difficult to compare them.

2. *Gradual optimization of individual criteria after the introduction of priorities*

or them with or without the allocation of concessions. This method is based on the process of ordering the criteria of importance and the procedures of building the sequence optimization, first on the first, then on the second, third, etc. The advantage of this method is a relatively high efficiency, when the extremes of the individual criteria are "gentle", then even with small values of concessions, a wide range of solutions is provided. The disadvantage of this method is the need to create an expert evaluation for setting priorities and awarding concessions, as well as the need to apply various optimization procedures.

3. *Optimization based on trade-offs introduced by assigning weights*

for each individual criterion or by assigning limit values for all but one of the criteria—the main criterion. The general criterion is formed by the sum of individual criteria with weighting factors. The advantage of this method—the decision on the quantity is also the Pareto-optimal solution.

4. *Optimization based on bringing the solution closer to a certain ideal value*

The first way to create an ideal value is to determine the optimal values of each individual criterion independently of the other criteria. Then find the minimum deviation from the ideal value. The second way to create an ideal value is to select the allowable values of each individual criterion as their value. Then look for a solution that is at least far from the ideal value. The disadvantage of this approach is that there is no information about the ideal value because many criteria are contradictory.

Compromise-based optimization is proposed when a general criterion is created as the sum of individual dimensionless criteria with weighting coefficients. The advantage of this method is simplicity and the solution is Pareto-optimal.

As an example of the selection of organizational and technical criteria, definition of the time interval is considered for the diagnosis of functional elements depending on the degree of responsibility of the mechatronic module, the node, the MS element and the rate of degradation processes. In the degradation process, it refers to depreciation

Table 3.8 Possibility of using diagnostic systems

Accident cost (destruction)	Slow rate of degradation object of diagnosis	Average rate of degradation object of diagnosis	High degree of degradation or sudden departure of the subject of diagnosis
Significant accident costs (destruction)	Portable diagnostic devices $K_{otv} = 0.1$, $K_{degr} = 0.1$	Portable diagnostic devices $K_{otv} = 0.1$, $K_{degr} = 0.5$	Stationary diagnostic devices $K_{otv} = 0.1$, $K_{degr} = 0.9$
Average cost of accident consequences (destruction)	Portable diagnostic devices $K_{otv} = 0.5$, $K_{degr} = 0.1$	Stationary diagnostic devices $K_{otv} = 0.5$, $K_{degr} = 0.5$	Continuous protection and diagnostics systems $K_{otv} = 0.5$, $K_{degr} = 0.9$
High accident costs (destruction)	Stationary diagnostic devices $K_{otv} = 0.9$, $K_{degr} = 0.1$	Continuous protection and diagnostics systems $K_{otv} = 0.9$, $K_{degr} = 0.5$	Continuous protection and diagnostics systems $K_{otv} = 0.9$, $K_{degr} = 0.9$

and element destruction, loss of precision, performance degradation, accumulation of damage, etc. Table 3.8 shows the feasibility of using diagnostic systems.

Table 3.8 shows the feasibility of using diagnostic systems.

The slow rate of degradation of the object of diagnosis is determined by the slow processes that cause damage over months, years. For example, such processes include wear of machine parts, stress relaxation, metal leakage (dimensional damage), corrosion.

The average rate of degradation of an object of diagnosis is determined by processes that cause damage in minutes, hours. Such processes include, for example, thermal processes, changes in cutting forces due to tool wear, wear and splitting tools.

High speed or sudden departure of the object of diagnosis is determined by rapid processes that cause damage in seconds, a fraction of a second. These processes include, for example, fluctuations during cutting, periodic changes in cutting forces, changes in frictional forces, and formation.

Table 3.9 shows the feasibility of using diagnostic systems and types of MS maintenance depending on the type of equipment.

The consequences of an accident (destruction) are the main factor determining the feasibility of the use, form, and content of the diagnostic system. Another factor is the downtime or availability of spare parts.

Table 3.9 Possibility of using diagnostic systems and types of MS maintenance

Type of device	Possibility of using MS diagnostic systems	Types of MS service
Auxiliary, duplicated, periodically used equipment	Diagnostic and periodic diagnostics are not required for portable diagnostic devices	Repairs
Relevant main equipment	Portable diagnostic equipment, stationary diagnostic systems	Service according to the technical condition
One highly responsible equipment	System of continuous protection and diagnostics	Maintenance, continuous protection

3.7 Conclusions

Analysis of existing methods, models and algorithms for diagnostics of technical systems. The classification of diagnostic algorithms according to the use of mathematical apparatus and mode of operation is considered. Five principles of intelligent diagnostic systems are proposed. The stages of solving problems of mechatronic systems diagnostics using a neural network are highlighted. The results of the analysis of neural network applications for machine node diagnostics are presented. Information process models have been developed that determine how diagnostics are organized and information process models for deciding on the technical condition of modules, nodes and elements of mechatronic systems based on artificial intelligence methods. Three models of organization of mechatronic systems diagnostics are proposed: parallel, sequential, combined. The use of diagnostic parameters is proposed: current force, voltage, power, temperature and temperature fields, vibroacoustic parameters, accuracy of spatial positions, rigidity, motion parameters, power parameters, time intervals. An algorithm for the diagnosis of mechatronic systems has been developed. A block diagram of a hybrid intelligent diagnostic system is proposed. A model of the neural network of fuzzy inference was developed, on the basis of which an algorithm for deciding on the technical condition of mechatronic systems is created. An example of a neural network for bearing diagnostics is considered.

Hardware models of systems for mechatronic objects diagnostics are developed depending on the method of diagnostics: parallel, sequential, combined. An algorithm for the construction of devices for mechatronic systems diagnostics is proposed. An algorithm was developed for the construction of systems for mechatronic systems diagnostics on the example of a CNC machine. A criterion is proposed for the calculation of diagnostic intervals, which is related to the degree of responsibility of the mechatronic module, node, element and the flow rate of degradation processes in it. A generalized criterion of optimality of equipment for diagnostics of mechatronic systems as functional economic, organizational, and technical criteria is proposed. Suggestions for the application of the system for the diagnosis of mechatronic objects depending on the type of device are presented.

References

1. Bartelmus, W.: Diagnostic information on gearbox condition for mechatronic systems. Trans. Inst. Meas. Control **25**(5), 451–465 (2003). ISSN: 0142–3312
2. Beard, R.V.: Failure accommodation in linear system through selfreorganization. Ph.D. thesis, MIT, MA, USA, 376 p. (1971)
3. Bencsik, A.L., Szabo, J.Z., Lendvay, M.: Maintenance and diagnostic in mechatronic systems. In: Szakal, A. (ed.) IEEE 9th International Conference on Computational Cybernetics (ICCC 2013), Tihany, Hungary, 08–10 July 2013, pp. 141–146 (2013). ISBN: 978-1-4799-0063-3
4. Bezák, P., Božek, P., Nikitin, Y.: Advanced robotic grasping system using deep learning. In: Procedia Engineering: The 6th International Conference on Modelling of Mechanical and Mechatronic Systems MMaMS 2014, High Tatras, Slovakia, 25–27 Nov 2014, vol. 96, pp. 10–20 (2014), online. ISSN: 1877-7058
5. Bílý, M., Holub, R.: The automatic production system design. Course literature, Martin (1995)
6. Božek, P.: Nonlinear modeling of the arm dynamics for the manipulators with reverse sensory validation. Int. J. Adv. Robot. Syst. **16**(6), 1–9 (2019). ISSN: 1729-8806
7. Cacko, P., Krenicky, T., Dobransky, J.: Impact of an excessive wear of bearing on the mechatronic devices. Appl. Mech. Mater. **460**, 99–106 (2014). ISSN: 1660-9336
8. Cauffriez, L., Grondel, S.: Real-time monitoring and diagnosis of a train door mechatronic system. In: IEEE/ASME 10th International Conference on Mechatronic and Embedded Systems and Applications (MESA 2014), Senigallia, Italy, 10–12 Sept 2014 (2014). ISBN: 978-1-4799-2280-2
9. Garstka, T.: Mechatronic system for technical diagnostic using Barkhausen effect. J. Vibroeng. **14**(1), 45–51 (2012). ISSN: 1392-8716
10. Gutierrez-Carvajal, R.E., de Melo, L.F., Rosario, J.M., Machado, J.A.T.: Condition-based diagnosis of mechatronic systems using a fractional calculus approach. Int. J. Syst. Sci. **47**(9), 2169–2177 (2016). ISSN: 0020-7721
11. Hannius, O., Ring, D., Karlsson, J.: Derivation of diagnostic requirements for a distributed UAV turbofan engine control system. In: Proceedings of the 51st ASME Turbo Expo 2006, Barcelona, Spain, 06–11 May 2006, vol. 2 (2006), pp. 469–477. ISBN: 0-7918-4237-1
12. Krenicky, T.: Implementation of virtual instrumentation for machinery monitoring. In: Scientific Papers: Operation and Diagnostics of Machines and Production Systems Operational States, vol. 4. RAM-Verlag, Lüdenscheid, pp. 5–8 (2011). ISBN: 978-3-942303-10-1
13. Krenický, T., Jacko, P.: Real-time monitoring of technical systems operation. Strojárstvo Extra **5**, 32/1–32/2 (2011)
14. Matveev, S.A., et al.: Diagnostic and monitoring system for technical condition of electromechanical section of thermal control systems in spacecraft. Int. J. Math. Eng. Manag. Sci. **5**(1), 181–192 (2020). ISSN: 2455-7749
15. Muller, T., Lange, K., Breuer, A., Krieger, O., Form, T.: Automatic and experience-based diagnostics using distributed data and neural networks. In: VDI BERICHTE 13th International Congress on Electronic Systems Vehicles, Baden Baden, Germany, 10–11 Oct 2007, vol. 2000, pp. 593–605 (2007). ISBN 978-3-18-092000-9
16. Murcinkova, Z., Krenicky, T.: Implementation of virtual instrumentation for multiparametric technical system monitoring. In: SGEM 2013: 13th International Multidisciplinary Science and Geoconference, Albena, Bulgaria, 16–22 June 2013, vol. 1. Sofia: STEF92 Technology, pp. 139–144 (2013). ISBN: 978-954-91818-9-0
17. Noonan, D.P., Vitiello, V., Shang, J.Z., Payne, C.J., Yang, G.Z.: A modular, mechatronic joint design for a flexible access platform for MIS. In: 2011 IEEE/RSJ International Conference on Intelligent Robots and Systems, San Francisco, USA, 25–30 Sept 2011, pp. 949–954 (2011). ISBN: 978-1-61284-455-8
18. Papakonstantinou, N., Proper, S., O'Halloran, B., Turner, I.Y.: A plant-wide and function-specific hierarchical functional fault detection and identification (HFFDI) system for multiple fault scenarios on complex systems. In: ASME International Design Engineering Technical

Conferences and Computers and Information in Engineering Conference, Boston, USA, 02–05 Aug 2015 (2016). ISBN: 978-0-7918-5705-2
19. Rognvaldsson, T., Panholzer, G., Byttner, S., Svensson, M.: A self-organized approach for unsupervised fault detection in multiple systems. In: 19th IEEE International Conference on Pattern Recognition (ICPR 2008), Tampa, USA, 08–11 Dec 2008, pp. 3775–3778 (2008). ISBN: 978-1-4244-2174-9
20. Rothe, S., Soffker, D.: Development of a state-related evaluation for diagnostic-oriented data filtering approach. In: Chang, F.K., Kopsaftopoulos, F. (eds.) 10th International Workshop on Structural Health Monitoring (IWSHM), Stanford, USA, 01–03 Sept 2015, Stanford Univ, pp. 593–600 (2015). ISBN: 978-1-60595-275-8
21. Selig, M., Shi, Z., Ball, A., Schmidt, K.: A modern diagnostic approach for automobile systems condition monitoring. In: Ball, A., Mishra, R., Gu, F., Rao, B.K.N. (eds.) 25th International Congress on Condition Monitoring and Diagnostic Engineering (COMADEM 2012), Huddersfield, GB, 18–20 June 2012. J. Phys. Conf. Ser. **364** (2012). ISSN: 1742-6588
22. Stepanov, P., Nikitin, Y.: Diagnostics of mechatronic systems on the basis of neural networks with high-performance data collection. In: Brezina, T., Jablonski, R. (eds.) 10th International Conference on Mechatronics, Brno, Czech Republic, 07–09 Oct 2013, pp. 433–440 (2014). ISBN: 978-3-319-02294-9
23. Syfert, M., Wnuk, P.: Signal processing in diaster system for simulation and diagnostic purposes. In: Brezina, T., Jablonski, R. (eds.) 10th International Conference on Mechatronics, Brno, Czech Republic, 07–09 Oct 2013, pp. 441–448 (2014). ISBN: 978-3-319-02294-9
24. Zidek, K., Hosovsky, A.: Wireless device based on MEMS sensors and bluetooth low energy (LE/Smart) technology for diagnostics of mechatronic systems. In: Fabian, S., Krenicky, T. (eds.) Design, Testing and Characteristics of Mechatronic Devices. Applied Mechanics and Materials, vol. 460, pp. 13–21 (2014). ISSN: 1660-9336

Conference on Computers and Information in Engineering Conference, Boston, USA, 02–05 Aug 2015 (2015). ISBN: 978-0-7918-5705 [illegible]
12. Rognvaldsson, T., Panholzer, G., Byttner, S., Svensson, M.: A self-organized approach for unsupervised fault detection in multiple systems. In: 2008 19th International Conference on Pattern Recognition (ICPR 2008), Tampa, USA, [illegible] 2008, pp. [illegible] (2008). ISBN: 978-1-4244-[illegible]
13. [illegible]
14. [illegible]
15. [illegible]
16. [illegible]
17. [illegible]

Appendix A
Example of a CNC Machine Diagnostics Program

```
#include "stdafx.h"
#include "KurszpT.h"
#include "KurszpTDlg.h"
#ifdef _DEBUG
#define new DEBUG_NEW
#endif

class CAboutDlg : public CDialog
{
public:
        CAboutDlg();
        enum { IDD = IDD_ABOUTBOX };
        protected:
        virtual void DoDataExchange(CDataExchange* pDX);//DDX/DDV support
        protected:
        DECLARE_MESSAGE_MAP()
};

CAboutDlg::CAboutDlg() : CDialog(CAboutDlg::IDD)
{
}
void CAboutDlg::DoDataExchange(CDataExchange* pDX)
{
        CDialog::DoDataExchange(pDX);
}
BEGIN_MESSAGE_MAP(CAboutDlg, CDialog)
END_MESSAGE_MAP()
```


P. Božek et al., *Diagnostics of Mechatronic Systems*, Studies in Systems, Decision and Control 345, https://doi.org/10.1007/978-3-030-67055-9

```
CKurszpTDlg::CKurszpTDlg(CWnd* pParent /*=NULL*/)
	: CDialog(CKurszpTDlg::IDD, pParent)
	, res1(0)
	, res2(0)
	, res3(0)
	, res4(0)
	, res5(0)
	, r11(0.9)
	, r12(0.1)
	, r13(0.1)
	, r14(0)
	, r15(0.1)
	, r16(0.3)
	, r21(0.4)
	, r22(0.3)
	, r23(0.3)
	, r24(0.1)
	, r25(0.2)
	, r26(0.7)
	, r31(0.1)
	, r32(0.3)
	, r33(0.8)
	, r34(0.5)
	, r35(0.1)
	, r36(0.4)
	, r41(0.2)
	, r42(0.1)
	, r43(0.3)
	, r44(0.6)
	, r45(0.9)
	, r46(0.1)
	, r51(0.5)
	, r52(0.1)
	, r53(0.1)
	, r54(0.8)
	, r55(0.2)
	, r56(0.6)
{
	m_hIcon = AfxGetApp()->LoadIcon(IDR_MAINFRAME);
}
```

```
void CKurszpTDlg::DoDataExchange(CDataExchange* pDX)
{
        CDialog::DoDataExchange(pDX);
        DDX_Text(pDX, IDC_EDIT31, res1);
        DDX_Text(pDX, IDC_EDIT32, res2);
        DDX_Text(pDX, IDC_EDIT33, res3);
        DDX_Text(pDX, IDC_EDIT34, res4);
        DDX_Text(pDX, IDC_EDIT35, res5);
        DDX_Text(pDX, IDC_EDIT1, r11);
        DDX_Text(pDX, IDC_EDIT2, r12);
        DDX_Text(pDX, IDC_EDIT3, r13);
        DDX_Text(pDX, IDC_EDIT4, r14);
        DDX_Text(pDX, IDC_EDIT5, r15);
        DDX_Text(pDX, IDC_EDIT6, r16);
        DDX_Text(pDX, IDC_EDIT7, r21);
        DDX_Text(pDX, IDC_EDIT8, r22);
        DDX_Text(pDX, IDC_EDIT9, r23);
        DDX_Text(pDX, IDC_EDIT10, r24);
        DDX_Text(pDX, IDC_EDIT11, r25);
        DDX_Text(pDX, IDC_EDIT12, r26);
        DDX_Text(pDX, IDC_EDIT13, r31);
        DDX_Text(pDX, IDC_EDIT14, r32);
        DDX_Text(pDX, IDC_EDIT15, r33);
        DDX_Text(pDX, IDC_EDIT16, r34);
        DDX_Text(pDX, IDC_EDIT17, r35);
        DDX_Text(pDX, IDC_EDIT18, r36);
        DDX_Text(pDX, IDC_EDIT19, r41);
        DDX_Text(pDX, IDC_EDIT20, r42);
        DDX_Text(pDX, IDC_EDIT21, r43);
        DDX_Text(pDX, IDC_EDIT22, r44);
        DDX_Text(pDX, IDC_EDIT23, r45);
        DDX_Text(pDX, IDC_EDIT24, r46);
        DDX_Text(pDX, IDC_EDIT25, r51);
        DDX_Text(pDX, IDC_EDIT26, r52);
        DDX_Text(pDX, IDC_EDIT27, r53);
        DDX_Text(pDX, IDC_EDIT28, r54);
        DDX_Text(pDX, IDC_EDIT29, r55);
        DDX_Text(pDX, IDC_EDIT30, r56);
}
```

```
BEGIN_MESSAGE_MAP(CKurszpTDlg, CDialog)
	ON_WM_SYSCOMMAND()
	ON_WM_PAINT()
	ON_WM_QUERYDRAGICON()
	//}}AFX_MSG_MAP
	ON_BN_CLICKED(IDOK, &CKurszpTDlg::OnBnClickedOk)
	ON_BN_CLICKED(IDC_BUTTON1, &CKurszpTDlg::OnBnClickedButton1)
END_MESSAGE_MAP()

BOOL CKurszpTDlg::OnInitDialog()
{
	CDialog::OnInitDialog();
	// Add "About..." menu item to system menu.
	// IDM_ABOUTBOX must be in the system command range.
	ASSERT((IDM_ABOUTBOX & 0xFFF0) == IDM_ABOUTBOX);
	ASSERT(IDM_ABOUTBOX < 0xF000);

	CMenu* pSysMenu = GetSystemMenu(FALSE);
	if (pSysMenu != NULL)
	{
		CString strAboutMenu;
		strAboutMenu.LoadString(IDS_ABOUTBOX);
		if (!strAboutMenu.IsEmpty())
		{
			pSysMenu->AppendMenu(MF_SEPARATOR);
			pSysMenu->AppendMenu(MF_STRING, IDM_ABOUTBOX,
strAboutMenu);
		}
	}
	// Set the icon for this dialog.  The framework does this automatically
	//  when the application's main window is not a dialog
```

```
        SetIcon(m_hIcon, TRUE);                 // Set big icon
        SetIcon(m_hIcon, FALSE);                // Set small icon

        // TODO: Add extra initialization here

        return TRUE;  // return TRUE  unless you set the focus to a control
}

void CKurszpTDlg::OnSysCommand(UINT nID, LPARAM lParam)
{
        if ((nID & 0xFFF0) == IDM_ABOUTBOX)
        {
                CAboutDlg dlgAbout;
                dlgAbout.DoModal();
        }
        else
        {
                CDialog::OnSysCommand(nID, lParam);
        }
}

void CKurszpTDlg::OnPaint()
{
        if (IsIconic())
        {
                CPaintDC dc(this); // device context for painting
                SendMessage(WM_ICONERASEBKGND,
reinterpret_cast<WPARAM>(dc.GetSafeHdc()), 0);
                // Center icon in client rectangle
                int cxIcon = GetSystemMetrics(SM_CXICON);
                int cyIcon = GetSystemMetrics(SM_CYICON);
                CRect rect;
                GetClientRect(&rect);
                int x = (rect.Width() - cxIcon + 1) / 2;
                int y = (rect.Height() - cyIcon + 1) / 2;
                // Draw the icon
        dc.DrawIcon(x, y, m_hIcon);
        }
        else
        {
                CDialog::OnPaint();
        }
}

HCURSOR CKurszpTDlg::OnQueryDragIcon()
{
        return static_cast<HCURSOR>(m_hIcon);
}
```

```
void CKurszpTDlg::OnBnClickedOk()
{
        // TODO: Add your control notification handler code here
        UpdateData(true);
        float mas[5],maxim,ee,v[5],max[5];
        int a,b,c,d,e,f,i,j;
        if(IsDlgButtonChecked(IDC_RADIO1))
        {
                mas[0]=r11;mas[1]=r21;mas[2]=r31;mas[3]=r41;mas[4]=r51;
            max[0]=r11;max[1]=r21;max[2]=r31;max[3]=r41;max[4]=r51;
                maxim=max[0];
                for(i=0;i<5;i++)
                {
                        if(max[i]>maxim)
                        {
                                maxim=max[i];
                        }
                }
                if(mas[0]==maxim)
                {
                        SetDlgItemText(IDC_EDIT36,_T("a1>="));
                        res1=maxim;
                }
                if(mas[0]<maxim)
                {
                        SetDlgItemText(IDC_EDIT36,_T("a1<="));
                        res1=mas[0];
                }
                if(mas[0]>maxim)
                {
                        SetDlgItemText(IDC_EDIT36,_T("a1="));
                        res1=mas[0];
                }
        max[0]=r12;max[1]=r22;max[2]=r32;max[3]=r42;max[4]=r52;
                maxim=max[0];
                for(i=0;i<5;i++)
                {
                        if(max[i]>maxim)
                        {
                                maxim=max[i];
                        }
                }
                if(mas[1]==maxim)
        {
                        SetDlgItemText(IDC_EDIT37,_T("a2>="));
                        res2=maxim;
                }
```

```
        if(mas[1]<maxim)
        {
                SetDlgItemText(IDC_EDIT37,_T("a2<="));
                res2=mas[1];
        }
        if(mas[1]>maxim)
        {
                SetDlgItemText(IDC_EDIT37,_T("a2="));
                res2=mas[1];
        }
    max[0]=r13;max[1]=r23;max[2]=r33;max[3]=r43;max[4]=r53;
        maxim=max[0];
        for(i=0;i<5;i++)
        {
                if(max[i]>maxim)
                {
                        maxim=max[i];
                }
        }
        if(mas[2]==maxim)
        {
                SetDlgItemText(IDC_EDIT38,_T("a3>="));
                res3=maxim;
        }
        if(mas[2]<maxim)
        {
                SetDlgItemText(IDC_EDIT38,_T("a3<="));
                res3=mas[2];
        }
        if(mas[2]>maxim)
        {
                SetDlgItemText(IDC_EDIT38,_T("a3="));
                res3=mas[2];
        }
    max[0]=r14;max[1]=r24;max[2]=r34;max[3]=r44;max[4]=r54;
        maxim=max[0];
        for(i=0;i<5;i++)
        {
                if(max[i]>maxim)
                {
                        maxim=max[i];
                }
        }
        if(mas[3]==maxim)
        {
                SetDlgItemText(IDC_EDIT39,_T("a4>="));
                res4=maxim;
        }
```

```
		if(mas[3]<maxim)
		{
			SetDlgItemText(IDC_EDIT39,_T("a4<="));
			res4=mas[3];
		}
		if(mas[3]>maxim)
		{
			SetDlgItemText(IDC_EDIT39,_T("a4="));
		res4=mas[3];
		}
	max[0]=r15;max[1]=r25;max[2]=r35;max[3]=r45;max[4]=r55;
		maxim=max[0];
		for(i=0;i<5;i++)
		{
			if(max[i]>maxim)
			{
				maxim=max[i];
			}
		}
		if(mas[4]==maxim)
		{
			SetDlgItemText(IDC_EDIT40,_T("a5>="));
			res5=maxim;
		}
		if(mas[4]<maxim)
		{
			SetDlgItemText(IDC_EDIT40,_T("a5<="));
			res5=mas[4];
		}
		if(mas[4]>maxim)
		{
			SetDlgItemText(IDC_EDIT40,_T("a5="));
			res5=mas[4];
		}
	}
	/////////Complete1//////////
	if(IsDlgButtonChecked(IDC_RADIO2))
	{
		mas[0]=r12;mas[1]=r22;mas[2]=r32;mas[3]=r42;mas[4]=r52;
	max[0]=r11;max[1]=r21;max[2]=r31;max[3]=r41;max[4]=r51;
		maxim=max[0];
		for(i=0;i<5;i++)
		{
			if(max[i]>maxim)
			{
				maxim=max[i];
			}
		}
```

```
        if(mas[0]==maxim)
        {
                SetDlgItemText(IDC_EDIT36,_T("a1>="));
                res1=maxim;
        }
        if(mas[0]<maxim)
        {
                SetDlgItemText(IDC_EDIT36,_T("a1<="));
                res1=mas[0];
        }
        if(mas[0]>maxim)
        {
                SetDlgItemText(IDC_EDIT36,_T("a1="));
                res1=mas[0];
        }
max[0]=r12;max[1]=r22;max[2]=r32;max[3]=r42;max[4]=r52;
        maxim=max[0];
        for(i=0;i<5;i++)
        {
                if(max[i]>maxim)
                {
                        maxim=max[i];
                }
        }
        ///////поиск интервала
        if(mas[1]==maxim)
        {
                SetDlgItemText(IDC_EDIT37,_T("a2>="));
                res2=maxim;
        }
        if(mas[1]<maxim)
        {
                SetDlgItemText(IDC_EDIT37,_T("a2<="));
                res2=mas[1];
        }
        if(mas[1]>maxim)
        {
                SetDlgItemText(IDC_EDIT37,_T("a2="));
                res2=mas[1];
        }
max[0]=r13;max[1]=r23;max[2]=r33;max[3]=r43;max[4]=r53;
        maxim=max[0];
        for(i=0;i<5;i++)
        {
                if(max[i]>maxim)
                {
                        maxim=max[i];
                }
```

```
		}
		///////поиск интервала
		if(mas[2]==maxim)
		{
			SetDlgItemText(IDC_EDIT38,_T("a3>="));
			res3=maxim;
		}
		if(mas[2]<maxim)
		{
			SetDlgItemText(IDC_EDIT38,_T("a3<="));
			res3=mas[2];
		}
		if(mas[2]>maxim)
		{
			SetDlgItemText(IDC_EDIT38,_T("a3="));
			res3=mas[2];
		}
	max[0]=r14;max[1]=r24;max[2]=r34;max[3]=r44;max[4]=r54;
		maxim=max[0];
		for(i=0;i<5;i++)
		{
			if(max[i]>maxim)
			{
				maxim=max[i];
			}
		}
		if(mas[3]==maxim)
		{
			SetDlgItemText(IDC_EDIT39,_T("a4>="));
			res4=maxim;
		}
		if(mas[3]<maxim)
		{
			SetDlgItemText(IDC_EDIT39,_T("a4<="));
			res4=mas[3];
		}
		if(mas[3]>maxim)
		{
			SetDlgItemText(IDC_EDIT39,_T("a4="));
			res4=mas[3];
		}
	max[0]=r15;max[1]=r25;max[2]=r35;max[3]=r45;max[4]=r55;
		maxim=max[0];
		for(i=0;i<5;i++)
		{
			if(max[i]>maxim)
			{
				maxim=max[i];
```

```
                }
            }
            if(mas[4]==maxim)
            {
                SetDlgItemText(IDC_EDIT40,_T("a5>="));
                res5=maxim;
            }
            if(mas[4]<maxim)
            {
                SetDlgItemText(IDC_EDIT40,_T("a5<="));
                res5=mas[4];
            }
            if(mas[4]>maxim)
            {
                SetDlgItemText(IDC_EDIT40,_T("a5="));
                res5=mas[4];
            }
        }
}
void CKurszpTDlg::OnBnClickedButton1()
{
        // TODO: Add your control notification handler code here
        MessageBox(_T("CNC tool monitoring program"));
}
```

Appendix B
Source Code of the Fuzzification Program of an Input Variable Using the Gaussian Curve Membership Function

```
{The fuzzing input variable using the Gaussian curve function}
input x;

{Use of the Gaussian curve function}
function GausFM(x,sigma,c)
        GausFM = exp(- (0.5*( x - c) ^ 2 ) / sigma ^ 2);
end;

{Parameter fields for calculating expressions}

var sigma[3], c[3];

{The dimensions of the fields [3] are equal to the number of the expression of the language
variable}
initialization
{Initialization was performed once at the beginning of the model} const N = 3;
        {N - number of terms}
        MinX = -1;
        MaxX = 1;
```


P. Božek et al., *Diagnostics of Mechatronic Systems*, Studies in Systems, Decision and Control 345, https://doi.org/10.1007/978-3-030-67055-9

```
 {MinX..MaxX - range of input variables}
  { Function usage parameter values for each term -
even range distribution }
        for (i = 1, N)
                begin
                sigma[i] = (MaxX - MinX)/(4*(N - 1));
                c[i] = MinX + (i-1)*(MaxX - MinX)/((N - 1));
                end;
end;

for (i = 1,N)
        begin
        y[i]= GausFM(x,sigma[i],c[i]);
        end;

output y [3]
{The output dimension y [3] is equal to the number of expressions
input variable N = 3}
```

Appendix C
Source Code of the Conclusions of the Accumulation Program Fuzzy Rules of Production and Output Variable Defuzzification

```
{1. Accumulation of conclusions of fuzzy production rules
  2. Variable output error}

{ Input parameters - conclusions of fuzzy rules }
input F_L,
          F_M,
          F_H;

{Calculation of the triangular tool function for x with parameters a, b, c}
function TriangleFM(x,a,b,c)
          if (x>a) and (x<c) then
                    begin
                    if (x<b) then TriangleFM = 1-b/(b-a) + x /(b-a);
                    if (x>=b)then TriangleFM = 1-b/(b-c) + x /(b-c);
                    end else
                    TriangleFM =  0;
end;
```


P. Božek et al., *Diagnostics of Mechatronic Systems*, Studies in Systems, Decision and Control 345, https://doi.org/10.1007/978-3-030-67055-9

```
var a[3], b[3], c[3], x;

initialization
{Initialization was performed once at the beginning of the model}
const N = 3;
    {N - number of terms}
    {MinX..MaxX - range of input variables }
    IntCount = 100;
    { number of partitions of the output functions range
    for numerical integration during dephasing }
    MinX = -1;
    MaxX = 1;
    { Set the parameters of the participation function
     of N terms}
        a[1]=-1;  b[1]=-0.7;  c[1]=-0.4;  {L}
        a[2]=-0.4; b[2]=0;    c[2]=0.4;   {M}
        a[3]=0.4;  b[3]=0.7;  c[3]=1;     {H}
        {the numerical integration step}
        dX=(MaxX-MinX)/IntCount;
end;

{Accumulation of fuzzy rules conclusions. For each term using the prob method
(the result of the product, the conclusion of fuzzy rules for the relevant function
accessories)}
```

```
function F_L_Res(x)
        F_L_Res = (F_L * TriangleFM(x, a[1], b[1], c[1]));
end;

function F_M_Res(x)
        F_M_Res = (F_M * TriangleFM(x, a[2], b[2], c[2]));
end;

function F_H_Res(x)
        F_H_Res = F_H *TriangleFM(x, a[3], b[3], c[3]);
end;

{Accumulation of all expressions in one function,
which has maximized values
pf the participation function}
function Acc(x);
        Acc = max(F_L_Res(x),F_M_Res(x));
        Acc = max(Acc,F_H_Res(x));
end;

{Finding the acquired center of gravity
curve using numerical integration} cm = 0;
m = 1e-14;
{low number to avoid division by zero}
L = 0;
R = 0;
L = Acc(MinX);

for (val = MinX,MaxX,dX)
        begin
        R = Acc(val+dX)
        cm = cm + dX*((val*L)+(val+dX)*R)/2;
        m = m + dX*(L+R)/2;
        L = R;
        end;

y=cm/m;
output y[1];
```

The manufacturer's authorised representative in the EU is Springer Nature Customer Service Centre GmbH, Europaplatz 3, 69115 Heidelberg, Germany. If you have any concerns regarding our products, please contact ProductSafety@springernature.com

Printed and bound by CPI Group (UK) Ltd, Croydon, CR0 4YY
28/09/2026
02224086-0014